Galileo
Publishing

GIZA-GENESIS

Volume 2

THE SPHINX REVEALED

As Earth approaches the "end time", December 21, 2012, what sacred messages do the Egyptian Pyramids and Sphinx map for humanity?

Howard Middleton-Jones
and
James M. Wilkie

A Galileo Publication

Tempe, Arizona

www.galileopublishing.com

A Galileo Publishing Publication
Tempe, Arizona, USA

Middleton-Jones, Howard and Wilkie, James M.
Giza-Genesis: Volume 2, The Sphinx Revealed

ISBN 0971809836

Front cover artwork is a reconstruction based on the original cloth map, *Passage System of the Great Pyramid of Gizeh in the land of Egypt*, (June-July, 1909) which was included in a rare run of the book, *The Great Pyramid and Passages*, by Edgar Morton.

Layout design, Gary Fisher, Cover layout, Irin Smith (Galileo Publishing); Cover layout, graphics and illustrations, Jeff Didham (Envisage Design Group); crop circle photos, Lucy Pringle

www.galileopublishing.com

Acknowledgements

The front cover is a reconstruction based on the original cloth map, Passage System of the Great Pyramid of Gizeh in the land of Egypt, (June-July 1909) which was included in a rare run of the book The Great Pyramid and Passages, by Edgar Morton.

Contents of this book are the result of many years of original research, reflected in the graphically displayed constructed models and illustrations. For these we are indebted to the assistance of Jeff Didham of Winnipeg (Envisage Design Group) Canada.

We would also like to thank Jan Merta, Ph.D., co-founder of Ambilac, and John Thorvaldson and Gail Young for their support and encouragement in the on-going research at Ambilac. In addition, we are grateful to Julie King, one of our more recent associate researchers from Australia, who has contributed a great deal toward the understanding of more recent events. The relevant section of her work is included in the Appendix.

The following people have continuously contributed towards our research, and to them we also offer our appreciation; Edmund Meadows of Viking Remote Viewing, David and Virginia Kingston, University of Life, Brian Stokes of Swansea Astronomical Society for assistance in constructing and checking the relevant star charts, and Kent Steadman of Cyberspace Orbit.

The Crop circle photographs were taken by Lucy Pringle, one of the most dedicated researchers, photographers and authors in this exciting field.

CONTENTS

List of Illustrations

Chapter 15

Chapter 17

INTRODUCTION

In our final chapter 15 of Volume 1 we left our readers with a pictorial overview, or taster, as to what to expect in this, volume 2. In a nutshell, if one can even attempt such a feat, in this edition of our Giza-Genesis series, we offer a full explanation and demonstration towards the complete layout of our Giza Plateau model.

As we explained in our previous book, this model includes the complete layout of the plateau complex, not just the three main pyramid structures but the structures that make up the whole of the Giza plateau. That is, the so-called "Queen's Pyramids", the causeways, the associated temples and the Sphinx. We will demonstrate that all mirroring of the Pyramid complex is generated from the apex of the Great Pyramid of Khufu, and how the model reflects our three primary star lineages, thus substantiating our very existence. In this model, the Great Pyramid represents our earth, the Pyramid of Khefre our sun and the Pyramid of Menkaure one of the stars' from our own sun's direct lineage, located within the Pleiadean star system.

Bearing this concept in mind, the full Plateau model also conveys to us how our DNA, Pi and Phi is represented, where Khufu is the center point, Khefre the Pi element and Menkaure the Phi element as they relate to Khufu. In addition, we demonstrate how the Sphinx is a reasonable representation of man in his duality, the head of the Sphinx rising above the body of the "beast", representing the "soul of man", Thus demonstrating that wherever a double helix is present in the DNA makeup, this duality is always present.

Following on from the above concept, we go into great detail showing how Christ is actually represented as the Pi, or Alpha, and Phi or Omega, where Omega is the life that breathes into us.

Captain Bruce Cathie, as we explained in Volume 1, identified three Global electro-magnetic grid systems that criss-crossed over our planet in a network, much like a web. We now go one further and not only confirm the identity of these grids, but their location and source and who actually constructed them aeons ago. We then demonstrate that the meeting of these grids in mid 2001 ultimately offers many opportunities and threats. The resulting effects of these actions and re-actions are investigated in some detail, which leads us into the research of these earth change events including that of the much talked about forthcoming magnetic pole shift.

One of the most misunderstood and possibly to some, most complex subject is that of the understanding of the "new" time lines. That is the relationship of earth's relative time as we know it, with our sun's relative zero point, and how in turn universal time is calculated. We explain this scenario with recent examples and how earth's relative zero time is connected to the galactic zero time, and in turn how they are influenced by the sun, and in turn its point of origin.

From this point, we then demonstrate the presence of multi dimensions which may assist in understanding our perception of how we perceive events and how they are later affected by the reaction of universal time.

The FULL Plateau layout demonstrates all of the above, and we explain these criteria by providing accurate and detailed computer generated diagrams and illustrations, based on our many years of research.

And finally, but by no means last; **Who actually constructed the Pyramid complex** at Giza? A question that has continuously intrigued and baffled researchers since recorded history. In this volume we demonstrate the real purpose behind the building of this complex, with evidence to show that the builders had a strong connection with the constellation of Eridanus, often referred to as the river Nile of the sky. Was the Giza acropolis with all its associated structures including the Sphinx, not only built as a marker to identify the origins of its architects to other passing alien travelers, but to show the three grids as related to earth? With the definitive evidence we have uncovered to date, we believe this to be so.

This in itself would provoke a whole series of books, and thus bear in mind that although heavily illustrated, again this volume is merely a succinct overview of part of our findings.

Thus we CONTINUE our journey.

CHAPTER 1

EARLY HISTORY OF THE GIZA NECROPOLIS

As one may have discovered, there has been a diverse array of material written on the Great Pyramid of Khufu and its associated and surrounding structures. However, in order to demonstrate the complexity of the many concepts and ideas involved arising from this unique necropolis, we offer the reader a succinct and thorough overview of the early history and exploration of this complex.

This essential background material will aid and arm the reader with the necessary "tools" while studying our own exploration and research as we forge through all barriers of time and space to unravel the biggest secret ever.

Illustration 1-1

In what we term "recent history" regarding the Giza necropolis, that is the last 150 years or so, much has been recorded, collated and discussed, but until now no definitive concept of what this unique structure is all about has ever been put forward. Even when remote sensing evolved in the 1970's and satellite imagery assisted in understanding many other ancient sites, this necropolis always hid its biggest secret…that is until now.

In addition, the recent work carried out by such organizations as the A.R.E (Association for Research and Enlightenment) has continuously been kept from the public eyes, adding more confusion and speculation. This in turn can be extremely frustrating for both the serious researcher and interested reader alike. Thus in this second of our series, we intend to blow open the doors wide with a succinct and full appraisal of the background research, culminating in our full findings explaining the FULL plateau model.

One of the earliest writers on Egypt and in particular on the Giza neccropolis was Herodotus, a 5th century BCE historian and geographer. He was born in Halicarnassus in 485 BCE and was exiled to Samos when he was very young. He commenced his journeys in 445 BCE travelling through Egypt, Syria, Babylon, and Macedonia studying their cultures and histories. In his multiple volume history book entitled "Histories", he dedicated one whole book entirely to the people, culture and land of Egypt.

Herodotus talks much in his histories of the construction of the Great Pyramid and other pyramids and underground labyrinths. However, we must bear in mind that he was obtaining information handed down over thousands of years, and in addition his dimensions and locations may be confused, such as the Great Pyramid versus Saqqara.

In the second book entitled EUTERPE he mentions the following;

> To make it took ten years, as I said- or rather to make the causeway, the works on the mound where the pyramid stands, and the underground chambers, which Cheops intended as vaults for his own use: these were built on a sort of island, surrounded by water introduced from the Nile by a canal. The Pyramid itself was twenty years in the building. It is a square, eight hundred each way, and the height the same, built entirely of polished stone fitted together with the utmost care. The stones of which it is composed are none of them less than thirty feet in length 2.2.1

> The Pyramid was built in steps, battlement-wise, as it is called, or, according to others altar wise. After laying the stones for the base, they raised the remaining stones to their places by means of machines formed of short wooden planks 2.125.1

The previous extracts are taken from the History of Herodotus translated by George Rawlinson.

It was not until the beginning of the 9th century of our era that Caliph Abdullah Al Mamun, son of the famous Harun Al-Rashid, celebrated in The Arabian Nights, became excited by the legend that the great Pyramid contained a secret chamber. One with rare treasures such as arms that would not rust, glass that could be bent without breaking, and maps of the celestial and terrestrial spheres.

Arab historians describe how Al Mamun collected a vast conglomeration of engineers, architects, builders and stone masons to batter his way into the great Pyramid, building fires close to its massive limestone blocks, dowsing them with cold vinegar, battering them with rams.

A hot stifling passage was burrowed for over a hundred feet into the body of the pyramid without success. About to give up, the diggers heard something heavy fall somewhere in the interior. Changing their angle of attack they broke into a dark and narrow passage three and half feet by four feet that sloped back up at an angle of 26 degrees to the hidden pivotal entrance. Then down about 365 slippery feet to an unfinished empty pit dug into solid limestone directly below the apex of the pyramid.

Disappointed, the Arab treasures hunters retraced their steps to where a large prismatic stone lay on the floor of the passage. It had clearly covered the end of a large rectangular red and black granite plug that filled what looked like another passage sloping into the body of the pyramid. Of this passage no mention had been made by classical authors.

Could Al Mamun have stumbled upon a secret kept since the original construction of the building?

Unable to chip or dislodge the granite plug that weighed several tons, the Arabs cut around the plug into the softer limestone, only to find three more similar granite plugs. Whoever had designed the passage wanted it definitely blocked. But the mystery of how the blocks were triggered into place, by whom and how the perpetrators escaped remains a mystery to this day.

Beyond the third granite plug came a series of limestone plugs, but these could be chiselled into pieces to reveal a passage sloping up at the same steep 26-degree angle. 150 feet up this ascending passage the Arabs were able to stand upright.

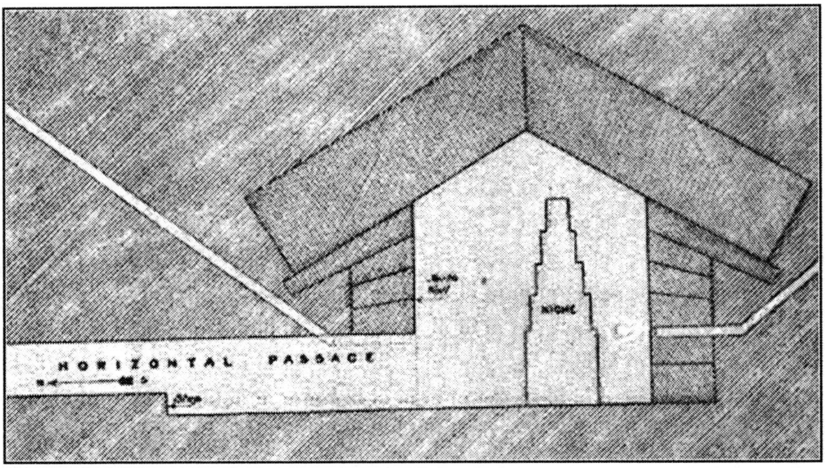

Illustration 1-2

Ahead stretched another low and narrow passage. It led to a bare room 18 feet long, almost square, with a niche in the left or eastern wall set into the limestone blocks, mysteriously coated with salt. Avid for treasure the Arabs hacked into the niche without coming to any door or treasure.

Because of their custom of placing women in tombs with gabled ceilings, the Arabs called this the Queens chamber. Crawling back they noted a strange step in the passage which neither they or anyone else until very recently could give an explanation.

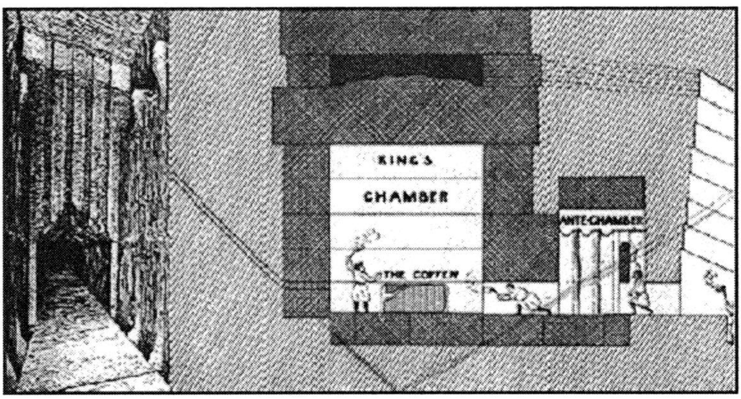

Illustration 1-3

Back where they could stand upright they raised their torches to reveal a Grand Gallery 28 feet in height that stretched up into the dark bowels of the pyramid at the same steep 26 degree slope. The narrow centre passage was extremely slippery, but slotted side ramps afforded a better foothold.

157 feet up the gallery they came to a large stone step three feet high that blocked the passage, forming a 6 by 8 foot platform. Its purpose as an indicator of a secret chamber has defied explanation by the experts until now.

For a further four centuries the great pile of limestone lay undisturbed, its outer casing intact except for the hole hacked by Al Mamun into its north face. Then a series of earthquakes demolished large parts of northern Egypt. To rebuild the city of Cairo with its splendid mosques, its inhabitants hacked away at the Pyramid's polished limestone casing to make building blocks.

In the course of several generations they managed to strip the entire 22 acres of 100 inch thick casing and expose a core of ascending and receding rougher limestone steps.

Around the stripped pyramids, fragments of rubble piled up so high they obliterated Al Mamun's entrance in the north face. However, removal of the outer casing brought to light two huge transoms embedded in the masonry, a protective gable over the tiny gaping original entrance to the descending passage.

Only now no one cared to enter the pyramid. Superstition shrouded the ancient structure. It was said to be alive with vermin, haunted by ghosts and by a naked woman with large teeth who seduced people into her power, driving them insane.

Illustration 1-4

Illustration 1-5

It was not until the beginning of the 17th century that an intrepid explorer entered the pyramid to measure its passages and chambers. John Greaves, a 36- year old English mathematician and astronomer, had come to Egypt hoping to find in the Great pyramid a datum that might help to establish the dimensions of the planet. The preceding century had spawned the great voyages of discovery, but no one yet knew the precise circumference of the earth. According to Galileo, then under house arrest for suggesting that the earth revolved around the sun instead of vice versa – a unit of measure could be found in the Great Pyramid accurately derived by ancient astronomers and geodesers from the actual circumference of the planet. Unfortunately, Greaves' measurement at the base of the Pyramid, covered as it was by much debris, failed to give him an accurate figure. For lack of this measure Sir Isaac Newton could not propound his general theory of gravitation, that all bodies in the universe attract each other in proportion to the product of their masses and inversely as the square of their distance apart. He had to wait until 1671 when a Frenchman, Jean Picard, successfully triangulated one degree of latitude across the French countryside from Dunkirk to Malvoisine.

With this great geodetic enterprise the Pyramid's earth commensurate values were forgotten. Its secrets remained as enigmatic as those of its neighbour the Sphinx, which by this time was almost obliterated by an accumulation of wind-blown sand from the Lybian desert.

Illustration 1-6

Only when young General Napoleon Bonaparte, bent on his conquest of Egypt, arrived at the Pyramid in 1798 with a following of so-called French "Savants" was the geodetic quest resumed.

Edme-Francois Jomard, one of the younger French savants, climbed up and down the rubble around the Pyramid to obtain a measure for one side of the base of 230.9 meters or 757 English feet. He then climbed to the top to obtain an apothem height of 184.722 meters. The apothem is the line from the peak from which a running drop of water will first reach the base. Jomard remembered that both Strabo and Diodorus Siculus, the Greek first century BC historian, had both reported that the apothem of the Pyramid measured precisely one stadium of 600 Greek feet, the same stadium from which our modern stadium is derived, and, more importantly, that this measure was accurately related to the size of the earth.

From further perusal of the classics Jomard knew that a stadium of 600 feet was considered 1/600 of a geographical degree. Calculating the geographical degree at the mean latitude of Egypt as 110.827.66 meters. Jomard divided by 600 to obtain 184.712 meters, again accurate to within 10 centimeters of his value of the Pyramid's apothem.

Could the builders of the Pyramid, Jomard wondered, have been able to work out their basic units of measure, stadium, cubit and foot from the size of the earth, and built this knowledge into the Pyramid?

Jomard's senior colleagues, classically indoctrinated, could not stomach the idea that their cherished Greeks might not be the founders of geometry. Re-measuring up and down the rubble they obtained a base two meters longer, enough to squelch young Jomard. His quest was dropped. Soured academicians mocked the theory that the ancient Egyptians could have possessed knowledge of geometry, geodesy and astronomy sufficient to validate Jomard's theories.

Illustration 1-7

As recently as 1963 an eminent engineer in Baltimore, author of an expensive privately printed booklet, "Designing and Building The Great Pyramid", claimed that "it is unlikely the Egyptians had more than a vague idea, if any, of the four cardinal points. There is no evidence in the Great Pyramid that they had any conception of true north or knew that a north-south line was perpendicular to an east-west line." In fact, the builders of the Pyramid knew true north so well that compasses are adjusted to its orientation not vice versa.

This was a well-known fact proved by the eminent British astronomer Richard A Proctor. As Proctor reviews the evidence the ancient builders could obtain a perfect north-south line by digging their descending passage into the rock base of the Pyramid using the circumpolar star Alpha Draconis whose rays at its lowest culmination, three degrees 43 minutes south of the Pole, would shine directly down a passage dug at the 30th parallel at an incline of 26 degrees 17 minutes.

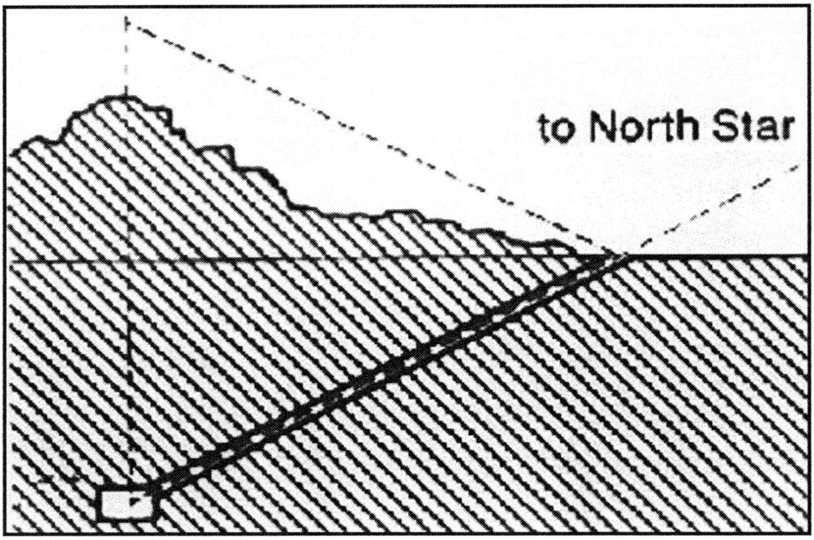

Illustration 1-8

Proctor's analysis provides an explanation for the quite extraordinary straightness of the walls of the Pyramid's descending passage, whose mean variation from a central axis along the entire length of 350 feet is less than ¼ inch in azimuth (from side to side) and only .1 of an inch in altitude (up and down). In that part nearest the aperture, which was to be the most important for celestial observation, the exactness is even greater, the mean error amounts to less than 1/50th of an inch.

Having measured the length of their descending passage and the angle of its descent, it would have been simple to locate a central spot with elementary trigonometry above the end of the descending passage and thus establish a centre for the proposed pyramid. Even if this were on rough elevated ground.

By continuing their tunnelling up through the lowest tiers of the growing pyramid, the buildings could maintain a precise orientation for at least the first ten courses, or until their tunnelling debauched from the growing pyramid.

To continue up the pyramid, the builders hit on the idea of creating an ascending passage at the reflecting angle of another 26 degrees 17 minutes. By plugging the descending passage and filling it with water or mercury they could reflect the polar star back up the ascending passage, truly aligned, as they built course by course. This is the very same system used today at the US Naval observatory in Washington DC where the daily transit of stars is noted to a split second by their reflection in a pool of mercury.

For the descending passage to hold water, says Proctor, its masonry at the point of juncture would have had to be hard rock, carefully joined. As it happens the stones at this point are quite different from the rest of the passage; harder, smoother and more finely jointed.

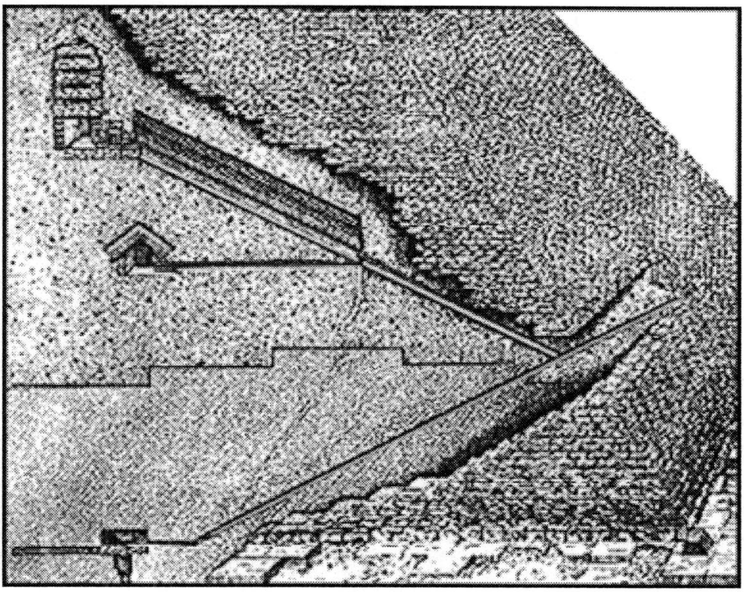

Illustration 1-9

For the ancients to create a firm body of astronomical data, they needed this perfectly aligned north passage to extrapolate a meridian across the heavenly vault. They could hardly hold up a string 500 feet high. They needed a slot through which to detect the precise movement of stars, sun, planets and moon as they transit the meridian in their apparent rotation through the heavens.

With the Grand Gallery the ancients has accomplished just such a feat.

The fact that the walls of the gallery are corbelled, with removable top stones made it possible to observe the northern sky as accurately as the southern. By placing observers not only in the gallery but outside, at cardinal points of the great truncated Pyramid the entire visible sky, says Proctor, could be accurately plotted. Even the apparently retrogressive paths of the planets could be noted.

To Proctor the Great Pyramid thus constructed would have been the greatest observatory and the most perfect until the development of the art of the telescope.

That the designers of the Pyramid aligned it to the stars is evident from two small shafts running north-south out of the Kings chamber deliberately orientated. What has taken much longer to realize is that other hidden shafts in the body of the building were deliberately incorporated as part of the original blueprint. The strongest evidence that the ancient Egyptians were capable of accurate astronomical observations comes from the fields of geodesy and geography, sciences whose subject is to determine the size and shape of the earth and locate landmarks.

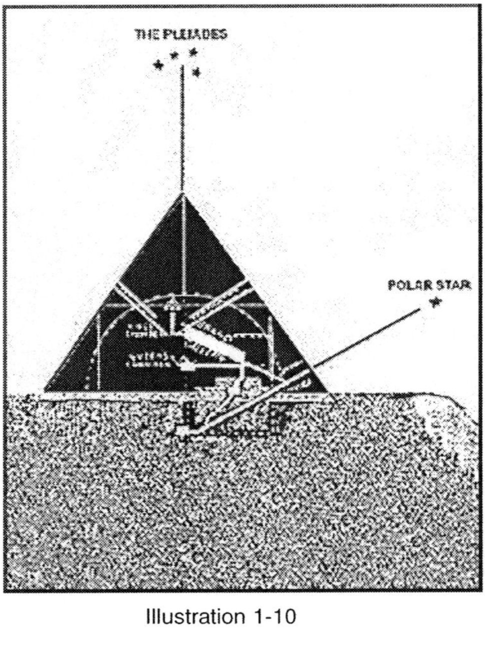

Illustration 1-10

Livio Catullo Stecchini, late Professor of the History of Science, shows that in the third millennium BC Egyptians could measure latitude to within a few hundred feet, an accomplishment not repeated on this planet until the eighteenth century of our era.

Ancient texts and hieroglyphs vindicate Jomard in full. The texts show that at least as early as the unification of Egypt, circa 2,800 BC the Egyptians knew the length of the circumference of the earth very precisely, the length of their country almost to the cubit, and the geographical coordinates of all the main points in their realm from the Equator to the Mediterranean. To do so the Egyptians must have been able to make astronomical observations with almost the exactness afforded by the modern telescope and chronometer.

The ancient texts make the length of Egypt from Behdet, a geodetic point on the Mediterranean at 31 degrees 30 minutes, exactly 1,800,000 of Jomard's cubits or 831,248 meters, all the way south to another geodetic point at Syene at 24 degrees of latitude.

Computing from texts dating to the third millennium BC the mean length of a degree of latitude in Egypt would be 110,832 meters. The modern estimate is 110,800 meters. The prime meridian of Egypt (running right through the Great Pyramid) was made to split the country longitudinally from Behdet on the Mediterranean to where it crosses the second cataract, exactly seven degrees to the south.

To simplify the dimensions of Northern Egypt, the Egyptian geographers accurately marked it as a triangle exactly one degree deep with its apex where the Nile splits just north of the Pyramid, fanning out one degree 24 minutes east and west to the sea. This became an actual delta shape whose angles were designed by the shadows cast by the northeast and northwest corners of the Great Pyramid.

Southern Egypt was made to run precisely six degrees to the first Cataracts of Aswan on the Tropic of Cancer. Two lines drawn parallel east and west of the prime meridian starting at the sides of the mouths of the Nile in the Delta, running to the Tropic, made a simple rectangle of Lower and Upper Egypt.

Professor Stecchini points out that when this particular geodetic system of ancient Egypt was established, the Tropic of Cancer was at 23 degrees 51 minutes that corresponds to a place on the Nile just south of the First Cataract. There a well was situated whose bottom was completely illuminated only one day a year when the sun stood directly overhead at the solstice.

But the ancients were sophisticated enough to know that the measurement had to be made from the edge not the centre of the sun, hence the well had to be placed 15 minutes north of the Tropic to account for the half diameter of the sun.

Cities and temples were then deliberately built at distances in round figures and simple fractions from the tropic or the prime meridian. As each of these geodetic centres was a political as well as a geographical "navel" of the world, an omphalos, or stone navel, was placed there to represent the northern hemisphere from equator to pole. It was marked out with meridians and parallels, showing the direction and distance of other such navels.

For the Egyptians to have laid out an absolutely straight meridian over 30 degrees of latitude from the Mediterranean to the equator covering 2000 miles, and drawn two more meridians equidistant east and west as boundaries of the country, must have required an enormous amount of personnel and careful astronomical sightings.

Even more sophisticated, as reconstructed by Stecchini, was their method of establishing longitude with an aid of an elementary system of telegraphy consisting of a series of beacons. The Egyptians were able to note what star was its zenith, directly overhead, at a certain moment, and flash the data via a string of flares to other observers so many degrees to east and west. Because every observable star comes to the meridian of every place on the globe once in 24 hours, the interval which elapses between the same star coming up on the meridian of two different places in the difference in longitude of the two places. Travelling further afield, the ancient geographer could establish his longitude with real precision on the basis of accurate tables of the nightly transit of celestial bodies as observed at the pyramid.

That the Great Pyramid was an accurate scale model of the northern hemisphere had to wait until 1925 for perfect confirmation. It was only then that the Egyptian government agreed to make an accurate survey on condition that Ludwig Borchardt and the German Institute of Archaeology in Cairo foot the bill for clearing all the rubble from the base of the Pyramid. J.H Cole was then employed to make a survey that ascertained to within millimeters the original position of the four corners.

If the base of the perimeter was intended to be 1/8 of a minute of degree, twice the perimeter will equal one minute. Cole's mean figure for twice the perimeter is 1,842.9 meters. The modern figure for a minute of latitude at the equator is almost exactly the same: 1,842,9 meters.

It is further evident that the ancient Egyptians knew that a degree of latitude is shortest at the equator and lengthens as it approaches the pole. The philosopher-Egyptologist R.A Schwaller de Lubicz noted that the Egyptians counted a minute of arc as being 1000 brasse or fathoms of six feet. According to Schwaller the brasse was a strictly geodetic measure that varied between 1.843 and 1.862 meters, depending upon the degree of latitude at which it was figured. The concordance is remarkable in that a minute of latitude is 1,842.9 meters at the equator and 1,861.65 at the pole.

It becomes clear that the tiny differences that appear in the foot in Persia, Greece, Mesopotamia and Egypt are due to the fact it was computed astronomically varying a meter fraction of a millimeter depending on the latitude at which it was measured. It is equally clear that the ancient world adopted a mathematical, geodetic, and astronomical system based on a circle of 360 degrees divided into 60 minutes and 60 seconds for a total of 1,296,000 seconds of arc.

To the Egyptians one second of arc consisted of 100 geographical feet. (Jomard's foot precisely of .308 meters). Six of these feet equalled one brasse or fathom; 600 feet, one stadium; 5\6000 feet, one mile; 60 miles, one degree of latitude; 360 degrees, the earth's circumference. Translated into meters 1,296,000 Egyptian feet are 39,916.8 Kilometers, which is one quarter of one percent off our modern earth circumference of 40,000 kilometers.

Even more subtlety, the ancients managed to achieve what Jefferson failed to accomplish for the Unites States when Congress was considering adopting the metric system. Jefferson wanted to include the element of time in a standard unit of measure, but was unable to do so, unaware that the Egyptians had long since resolved the problem with their geographic cubit.

129,600,000 Egyptian feet is equal to 86,400,000 geographical cubits (a cubit being a foot and a half) and that is exactly the number of seconds in a day of 24 hours. The Egyptians accurately estimated that the earth covers 1000 cubits in a second of time at the equator as it spins on its axis.

That Jomard's cubit of .462 meters is incorporated 500 times exactly in the base of the Great Pyramid and 400 times exactly in the apothem is no longer a matter of debate. Thanks to Cole it is simply a matter of multiplication. Even the confusing report that the base of the Pyramid measured 833 1/3 feet of the Roman historian Pliny becomes easily resolvable. The incongruity is rational when it is realized that Pliny's foot was 9/10 of a geographical foot. In other words 833 1/3 of Pliny's feet turn out to be 750 Egyptian geographic feet, or 500 geographical cubits.

So how did the ancients compute the circumference of the earth? Basically with the aid of two tall obelisks placed on the same meridian a considerable distance apart. By measuring the difference in the shadow of the two obelisks at the moment of the solstice they could extrapolate the circumference of the earth from the distance between the obelisks.

To measure accurately a large distance across rough terrain, the ancient geodesers developed a simple system, that of carrier pigeons. Knowing the speed of their pigeons they could deduce the distance they travelled in a certain time. This would give a fair answer. Repeat the process with a hundred or thousand pigeons and you obtain a very good circumference of the earth. That this was the Egyptian method is attested to by the Egyptian glyph for measuring latitude, two doves standing on an omphalos, holding a string in their beaks.

Note

The majority of the information above has been edited from the CD "Probing The Pyramids" © Ambilac Corporation; written by Peter Thomkins and James M Wilkie.

(Endnotes)

[1] Artaba. A unit of measurement consisting of 90 librae, which metrologists call a artaba, the Persian name. It was adapted as the official standard by the Persian Empire and its use under this Persian name became general throughout the ancient world. The artaba has the following contents;

29,160 cubic centimeters or grams
90 Roman Librae
1080 Roman ounces or 27 cubic centimeters
3200 Egyptian qedet of 9.1125 grams (Which equals 450,000 English grains)

The artaba was of the utmost importance in ancient Egypt because it was the standard ration of wheat for a month. That is, the ration for an adult free male, woman, slaves and was also the standard monthly ration in China.

CHAPTER 2

GIZA PLATEAU MODEL

As one can well imagine, with the extensive and diverse research that has taken place on the plateau at Giza, very little to this day is known about the people and culture who actually constructed the complex. Also, the great unknown, the reason why and the purpose for which the plateau is laid out, thus awaiting decipherment from the structures on and around the configuration located on the plateau itself.

However, what we have learned to this point in time is that there are very precise mathematical constants embedded in the overall model. These constants are incorporated into the measurements of each structure, and even more amazing is the fact that these constants are incorporated in the overall layout of ALL the structures located on the plateau. It is the consistency of these constants that have been incorporated into the structures that have brought forth a seemingly endless stream of theories as to the "hidden" meaning of those who blueprinted the complex itself. All of these proposed theories have consistently fallen short in explaining the purpose and the data for which it was intended to teach us, as humans, as our civilization progressed over time.

However, one theory that we will discuss in full in chapter three, is the possibility of the necropolis being a "marker" used to identify this planet to other passing alien travellers.

One would surmise that we should have been able to decipher this message that is truly written in stone, however, to this point in time all attempts have failed. The question then arose "what piece of information were we missing, or overlooking?" it was within that one single question that the answer was to be found. In Volume 1 we showed how the location to the Hall of Records was located by applying the system through the extension of alignments etc.

We then further introduced the "mirroring" concept as part of the required process. Then we introduced the "as above-so below" concept, and the three lineages were also covered. Now in Volume 2 we will explain the method for which the plateau should be drawn out. Along with this process that we will demonstrate, comes the understanding as to why it would have been impossible to construct it into a physical form, as this is a 3 Dimensional Model. In addition, we will then illustrate the evidence that those who built the structures did indeed leave behind their mark, and this mark points directly to their identity as well as their homeland.

Completing the Giza Overall Modeling

In volume 1 we illustrated the method to which mirroring was used to help read the plateau, the numbers used were those which repeatedly show up throughout the entire complex. These numbers are 3.71428, 3.142857, 26.42857 and 51.42857.

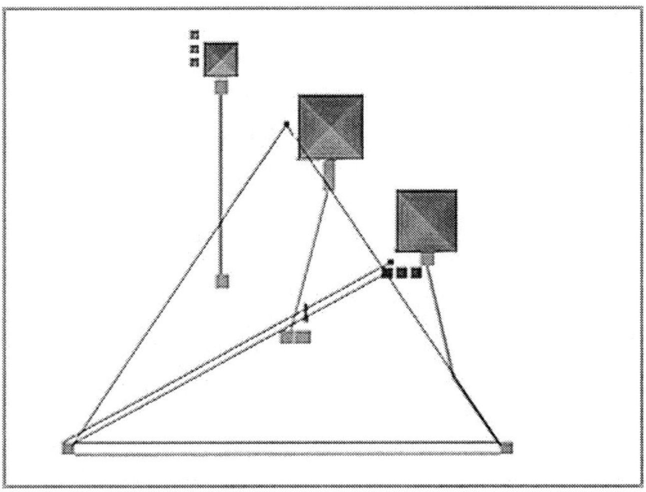

Illustration 2-1 Plateau Model No4 (Volume 1)

The above diagram also led us to the location of the Hall of Records, which we established at Latitude 29deg.58'29" Longitude 31deg.08'28". It is important to note that at the time the Giza complex was built that this location lay beneath the River Nile, while over time the Nile has since changed from its original path.

In volume 1 we also introduced the lineage model and disclosed the precision of alignments and how the alignments controlled (or effected maximum. / minimum control) over associated astral bodies.

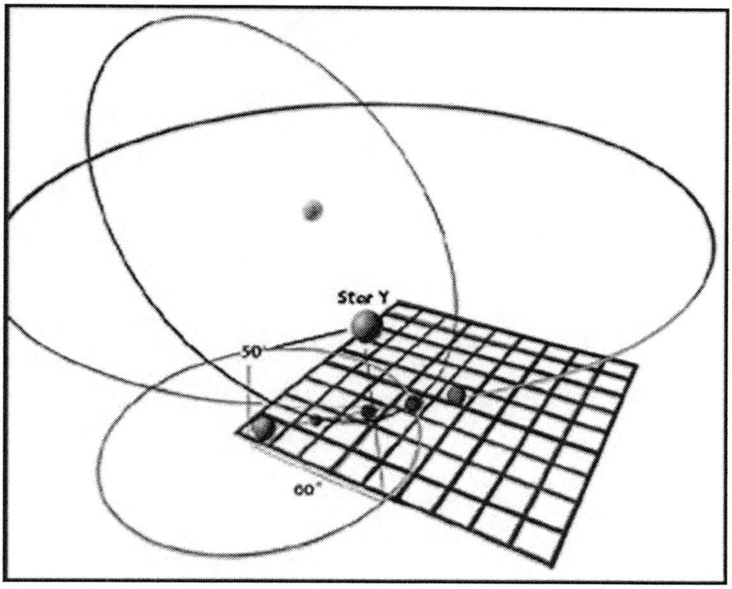

Illustration 2-2 solar grid No 3 (Volume 1)

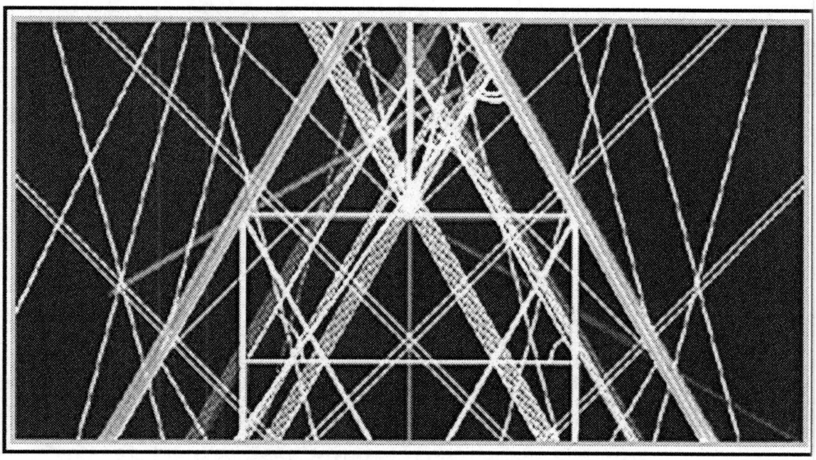

Illustration 2-3 Autocad model No2 (Volume 1)

In the latter portion of our first book showed how through the use of the mirroring and lineage models, we were able to establish, and consequently allow us to gain access to the complex as a result of this knowledge. The important point however, was that both access points resulted from that one single aspect of the much larger mirroring used to fully decipher the plateau.

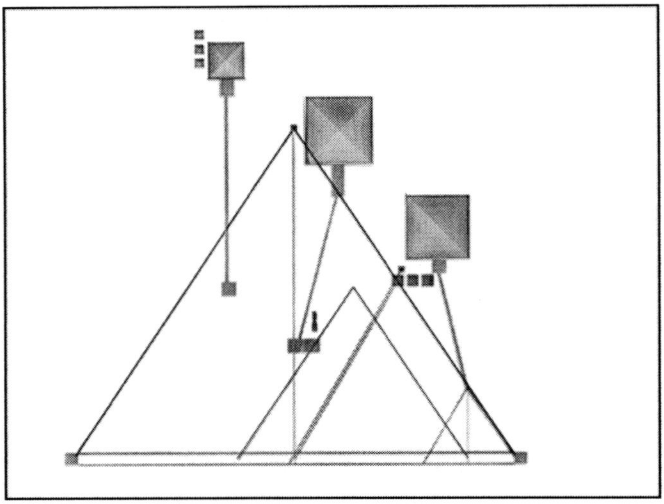

Illustration 2-4 Plateau model – main frame No 4 (Volume 1)

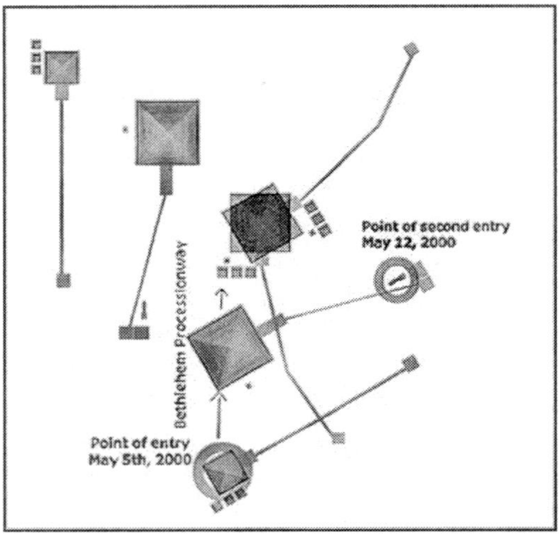

Illustration 2-5 Full Plateau model 120 degrees rotation

We will now describe the plateau model in full taking each construct step by step as it relates to this particular concept.

Knowing that there were indeed three lineages involved, having already deciphered that aspect from the present layout on the plateau, we soon realized that what was being shown was but one half of one lineage. Thus, through a series of mirroring we were able to fully construct the two-Dimensional model. However, bear in mind that not only did the model have to be mirrored but once completed, the entire model had to display itself mirrored. In other words, it had to reflect the adage, "As above so below".

Initially, we commenced with all the known Pyramids and their associated walkways with their attached temples including the Sphinx and its temple. The complex was then mirrored as a whole using the Pyramid of Khufu as the centre point.

2.1 Lineage- This mirroring continued as previously explained in Volume 1 of our work, which eventually produced the alignment of the star Alnitac in the constellation of Orion. Remember, that this star is precisely 51.4285 degrees in elevation over that point in the sky on December 21st 2012 at 10:18 and 13 seconds PM, local Cairo time, when viewed from the top of the satellite pyramid of Khufu we refer to as Horus.

2.2 Lineage- Further mirroring of the model produced our third lineage line, where it was found to be set precisely to the earlier alignment deciphered previously. This alignment was then used to calculate the portal alignment that took place on May 5th 2000, where we explained in full in volume 1.

2.3 Lineage- In Volume 3 of this series, we will go into greater detail and explain the deciphered data regarding the Belt of Orion, in that this could be utilized to orientate one's position in relationship to the remaining galaxies' more complex grid structure. The next stage was completed by mirroring the entire plateau model from the "0" degree base on which the pyramids were constructed. Having then completed this aspect of the modelling, not only did it reflect the "as above so below" concept, we noticed that the Pi and Phi dimensions incorporated into the sides of the Pyramid, placed the Pi portion aligning towards the centre pivot point. Thus leaving the Phi facing outward towards the outside perimeter!

Full Model It was at this juncture, after realizing we had produced an overview in Two-Dimensional form of an intended three-Dimensional model, we began to incorporate all the remaining data we had discovered in the original Plateau layout. For this, we focused our attention on the model's primary focal point, that being the Pyramid of Khufu. Then, we worked with the numbers and associated angles, those being; 50 plus 51.42857 degrees and 25 plus 26.42857 degrees. This latter angle being the angle at which the descending and ascending passages are set to inside the Pyramid of Khufu. It is true to say that this is not actually the angle they are measured at, unless you incorporate the "0" point from the outside edge of the corner casing stone as opposed to the midpoint. All of which is explained in detail in Volume 1.

As our lineage lines are processional and are linear, we find that the alignments are, to each lineage line, in a straight line that must be set at its true angle. These angles must also take in account for the value of "0" relative. In addition, while knowing that the 50 degree angle "plus" the value of "0" was placing the two outside lineages at 51.42857 angles, the question posed was which of the lineages were elevated 51.42857 degrees above the horizon? Also, which end of the respective lineages were angled down below the horizon?

This is when the true beauty of the choice of which set to construct became so profound, as the three Pyramids were actually chosen to reflect the three stars as they appear in the sky of the belt of Orion. Having already previously established from our earlier work with the Plateau model we knew that the Orion connection to its lineage was on a different set. In addition, we also established (as explained in Volume 1) that at the time of alignment the star Alnitac would have been at a very precise 51.42857 degree elevation above the horizon when viewed from the Plateau.

As the existing set present on the Plateau was also visible above the base line, we may fully conclude that this set indicated that on this lineage line it was the elevated side of the other intersecting lineage.

Thus, having established the positioning of two of the three lineage lines, it now left us with the question what is the placement angle of the third set? Upon careful review, we concluded that the angle when intersected at precisely 30 degrees was indeed the correct placement point of this intersection. It must be borne in mind that the angles of the descending and ascending descending passageways located in the interior of the Pyramid of Khufu once corrected, as we pointed out in Volume 1, was the true angle of the lineage line. Thus the placement of the entrance of Khufu on the north face is slightly offset from centre, this angle being 26.42857 – 1.42857 = 25 degrees. Remember, the floor level of the Queens Chamber[1] rests on the 25th course of the Pyramid itself, and the passageway to the chamber is at 1.42857 degrees above that level, as explained in Volume 1 and expanded upon in chapter 4 in this volume.

This was then integrated into our present model. Hence, the alignments from the exterior, aligning to the three girdle stones, located in the ascending passage, become very important to note, as does the reason for the moving plugs which they held apart. This aspect is explained in detail in a later chapter in this volume dealing with the "Grids".

(Endnotes)

[1] The down angle of the "air shaft" located in the Queens Chamber is set at precisely 45 degrees plus the value of zero, while the second "up" angle shaft is set at 45 degrees, and which is "out going". The "zero" point being represented by the figure 10 (the 1 or element in question added to its first zero point) which is divided to its centre point, represented as 5. We then add this to our respective shaft angles and once again come to our 50 and 51.42857 degrees. Bear I mind that the "niche" in the Queens Chamber is inset where the widest end is at floor level. Thus the floor level being 254 and the angle of the relative "zero" of earth at 5, give us our precise 30degree angle. This is the reason the intersection of our ascending and descending passageways are placed on the fifth course of the Pyramid.

CHAPTER 3
THE ERIDANUS CONNECTION

Next to the shear massive size of the Pyramid Khufu the most impressive feature found on the plateau is the Sphinx. In Arabic this is known as Abu'l-hawl, the wonderful, or terrible one. The sphinx itself has been drawing much attention even as recently as the last few years, where there is an attempt to put an age to the approximate time of its construction. While we would not like to even theorize as to the date of construction at this time, we certainly would like to point out the reason the structure is where it is and why it was placed on the plateau in the first place. YES this subject has been debated quite fiercely in the sometimes-shallow world of academia.

The Sphinx has been cleared by various people over time, the earliest recorded being that of the French engineers of Napoleon's expedition of 1799. In 1816 Captain Caviglia excavated the Sphinx , where he noted the double casing on the body and paws and the remains of red pigment with which one side had been painted. After shifting the great mounds of sand that covered the monument, he discovered the stela of Thutmosis IV.

By the 18th Dynasty, the small 4th Dynasty (traditionally believed to date) temple in front of the Sphinx was totally covered and had been forgotten about, while in the Roman period , a stairway and ramp were built over the whole structure.

Mariette failed to clear it when dealing with the valley temple in the 1860's and thereafter the work continued by Maspero. But he also never finished the work.

In 1925 Baraize was entrusted with the task of clearing away the sand that once more had covered the Sphinx, and he built large coffer-like walls, which were demolished by Selim Hassan in 1936-37.

During the Middle Kingdom (2050-1786 BCE) along with other sphinxes, it was known as seshep-ankh, the living statue, and by the New Kingdom (1567-1085 BCE) was associated with the sun god and regarded as a version of Hor-em-akhet, or Hor-akhty (Horus of the Horizon). This seemed an appropriate name as the whole cemetery was called Akhet Khufu, or the Horizon of Khufu.

During the Ptolemaic (323-30 BCE) period the Sphinx must have been freed from the sand as a poem was discovered scratched, in Greek, on its toes; in part it reads:

Our hands hold not the scimitar
But the fraternal cup of the banquet
And all night long while the sacrifices are burning
We sing songs to Harmakis (Hor-em-Akhet)
And our heads are decorated with garlands
From Selim Hassan, *The Sphinx and its secrets*

The important point to note is that even with the recent and on-going debates on the waterline dating of theSphinx, the actual dating for the construction in this manner will never be established. There is in fact no evidence to show whether or not the Sphinx had a much harder casing stone that encased the structure at the time of construction. If indeed this was the case, it most certainly would have afforded much added protection from the climatic weathering, and if so the dating could possibly go way back hundreds of thousands of years.

Before we could come forth with our findings many things had to be addressed, such as the Sphinx temple and the placement of the Sphinx temple being located directly beside the Valley Temple of Khefre. Next was the Egyptian mythology behind the Sun god Re (or Ra).

However, these concerns were only secondary to the big question... is it time?

We believe that it truly is time for bringing this information forward, with caution and warning, for Truth is stronger (and often stranger) than fiction.

The complex was built with the labour of our "primitive" ancestors, who for a lack of another word were used and held as slaves to this massive construction project. These people were forced to work under direction from an alien race known as the Erideans. The Erideans are a highly technologically advanced race having holdings on many planets located around many of the Suns situated within our part of the Galaxy. The Earth at the time of the construction had been mined by this race for centuries previous to this time in history. For they were truly the owners and landlords of that property in and around Giza, and to this day as far as we are aware, still own rights to it, that is except for a very small portion of space. This in fact was the field where a cave was situated on the perimeter, the name of this well or cave was called Machpelah.

At Abraham's insistence Ephron sold to Abraham (in the presence of Ephrons' council) this property for the sum of 400 shekels of silver. (Genesis 23. 1-20) Bear in mind that Abraham was considered a resident alien whilst he was among the Hittites. The Hittites are from the Eridanus [1] constellation close to the star Rana. (Lt. Frog). Later, we will disclose more on this cave as well as its significance and why Abraham insisted that he pay for it in front of the Hittite council, thus making the purchase a witnessed record of the transaction.

The Constellation Eridanus is seen to meander from the northern to the southern sky, and in past times it was difficult to identify which river the constellation represents; some writers claimed it was the Tigris or Euphrates, others the Nile. While Homer referred to it as an "ocean stream", in fact it may originally have been a mythic river, and not meant to be identified with any geographical site. Eridanus might represent the Sumerian Strong River, also known as Ariadan. Or it might be one of more of any other important rivers—the Ganges, Po, Euphrates, or Nile. However, we suggest that in fact Eridanus has very strong ties not only with the River Nile but with the actual complete layout of the Giza Necropolis itself.

In the north, peoples could see it in general, but not all of the year. In the south, however, it was visible throughout the year by the ancient astronomers. This constellation was one of the most important of the constellations which formed the most ancient Upper Egyptian religion and mythical traditions of the Osirian (Weserian) beliefs as well as the Lower Egyptian Sun religion In turn this was amalgamated by each at a very early pre-dynastic, probably pre-historic time.

The stars in the Orion Constellation (Rigel, and Sirius, as an example) were always associated with the Eridanian constellation. This is the 'heavenly region' where the kings were believed to travel and reside after death, and often associated with many 'sacred directives' in order to accomplish this trek.

Stories that are associated with these constellations include the later Greek tale of Phaethon, son of Helios the sun god, who while driving the cart containing the sun, came too near Ursa Major, the warming made the 'bear' attempt to find 'refreshment' in a sea. In doing this he awoke Draco, the dragon, who frightened Bootes Bootis being the brightest star of the constellation, Arcturus —Greek "Keeper of the Bear".. Bootis neared the constellation Scorpio (the scorpion), and fearing that Bootes would be bitten by Scorpions deadly tail, Phaethon dropped the reigns of the horses drawing the cart of the sun. It headed nearer to Earth, causing great droughts, drying up many seas, and turning the Sahara into a desert. Zeus struck Phaethon with a thunderbolt, making him fall into Eridanus and drown. Could this be an analogy of some type of astronomic event which may have caused a catastrophic event on Earth?

One must understand that the Eridean people who were mining this planet had in fact been mining it for some time, long before any other alien races were visiting this planet. Almost all the labyrinth of tunnels running beneath the Earths' surface are a result of Eridean mining, and as we stated this process began long before this planet was stabilized. For in fact, it was the Erideans themselves who towed our Moon into place and consequently set it into its present orbit. In addition, while this was being carried out they set up the crystal grid below the surface of the planet in order to stabilize the planet and prepare it for surface habitation. This is the main reason that the crystal grid is constructed using Giza as its focal point. It was much later after the Erideans had stabilized the planet that the Anunnaki eventually showed up here on this planet. We do not know the cause or events that did actually occur however we can emphatically state that the Erideans are not friends of the Anunnaki!

Much work has been published in recent years on the subject of the Anunnaki and their culture, Sitchin [2] one of the rare researchers who has intimate knowledge of the Babylonian cultures, has written much on this subject. While, in addition, David Icke a well known controversial journalist and speaker, has also written widely on this subject.

The only suggestion we would make at this time regarding the works of Sitchin and Icke, would be in Sitchins reference to the 12th Planet. If one were to regard this planet, as by definition, as Niburu as Sitchin suggests, then we offer the premise that this may in fact not be a returning planet, but a newly created one. It could be of tremendous value if Sitchin could re-evaluate his part as it relates to his research that may help to confirm the findings in either case.

We strongly believe, therefore, that the interpretation should have been the 12th Star. However, if indeed this Niburu is a returning planet, it is not one that has inhabited this solar system previously, but at this time we do not have sufficient data to comment further on this aspect. But we can say from our research that if it is the birth of a planet, then this will occur on the 21st December 2001 from within our sun. This will arise as a mass exiting through our sun, where this mass itself is the product of a parallel universe that is also undergoing an advancement in transition and grid shifts. For it is the sun's interdimensional and interlocking gridlines that create an opportunity for a portal to open through these grid lines, thus enabling this planetary mass ejection to occur. According to our calculations the grid alignment that occurred that initiated this process occurred on September 13th 2000, as we stated on the Internet at that time. For more details on this event, please refer to chapters 13 and 14. For more details on this please refer to chapter 4 The Grids, and Volume 3. In addition, we would like to comment that at present we have insufficient data to fully justify this thesis, however, the available research concurs this to be a concept of a parallel universe.

This 12th Star would be then visible at the time of the full alignment that will occur on the 21st of December 2012. For it will only be an alignment in full once the 12 Suns are in their exact locations that this can occur. Therefore the closer we move towards that date the closer we move towards the alignment. As our model indicates (as deciphered from the plateau) there are 12 Suns that will be in the alignment at the end time of December 21st 2012.

When the planet had been stabilized, only then did visitors from many other worlds begin to visit this planet we now call Earth. To identify this planet as a holding of the Erideans the plateau site was then marked thus making those whom visited aware that it was in fact an Eridean holding. Although we have no way to confirm that all inhabitable planets are marked or identified elsewhere in the Galaxy, as to whom owns (or claims ownership or possession) each planet, for we do not possess the technology at present required to view or visit any planets outside of our own Solar System. However, based on the information at hand we would like to point out the relative structures (under great debate) discovered on the planet Mars, in the area known as Cydonia.

Note that reports of structures located in and around the structure referred to as the face of Cydonia are placed in symmetrical pattern, very similar in fashion to those located on the Giza Plateau. It is also our contention that at the time Mars was at some stage an Eridean holding and, like the Earth had once been controlled by an artificial grid (crystal grid) set up by the Erideans. Phobos (presently on an unstable orbit) had also been towed into place just like our moon was towed into place at a much later date in time.

These moons are required in order to maintain stability once the planet is Terra-formed. Terra forming is only possible once the planet has been itself first stabilized, by this we mean that the artificial grid must first be put into place, thus stabilizing the electro magnetic flow within itself, to be then set or adjusted according to its placement within the outer incoming magnetic fields, with which it would be in constant interaction.

By the time the artificial grid that was established on Mars in its early days had maximized its settings, the Earth was already in the final stages of its artificial grid activation process. Once End Time was reached for the then terra formed Mars, Earths grid was activated and terra forming and re-population began.

We now have seen satellite imagery taken of the planet Venus that would indicate that the process of laying down an interior grid (crystal grid) has been underway for some time judging by the structures that are visible at the tunnel entrances. BUT who is responsible for this work in progress?

IF the Erideans are responsible for this terra-forming in this Solar System how would we know or be able to prove it. SIMPLE, they left their address!!

On the plateau we find the Sphinx located on the west side directly behind the Sphinx Sun Temple, which is located adjacent to the Valley Temple of Khafre. As we pointed out in Volume 1, the North and South walls of the respective temples fit side by side with angled walls. This was the method that was used to locate the Hall of Records that we established as being at Latitude 29deg. 58’29” Longitude 31deg.08’28.

To venture to that location looking for a chamber containing Records is of course absurd. The whole purpose of the exercise was to establish a location from which all the information contained in the finished (deciphered) plateau model was known, could be precisely read. Although Edgar Casey gave some very accurate readings on the subject it was one’s own inability that prohibited anyone from prematurely deciphering the same. For he did point out that the hall of records could be opened as early as 19??, but would be open in 1998. And we are happy to inform you that in 1998 it was. Casey’s second clue was that the Star Polaris would appear to be out of place, which it was found to be until the correct time was calculated into the data. That date turns out to be Dec. 21st, 2012 or commonly referred to by the Mayans as the ‘End Time”.

From our position we identified as the Hall of Records, we began to check the fully deciphered plateau model against the location of the stars in the sky at that time. Once we begin to align the stars against the information contained in the plateau model, we notice that due East of the Sphinx at precisely 15 degrees below the horizon sits the Star known as Rana. It is true that Rana will be at various altitudes and will pass across the front of the Sphinx many times etc. in the skies due East. However, it is shown by the precision placement of the Sphinx that at the time of the alignment it MUST be precisely 15 degrees below the zero elevation line, or what equates to being called the horizon. This data is derived from the following observations:

> First, the Pyramids are very precisely built on top of an extremely well leveled- out base height, this height is held constant.
>
> Second, the base of the Sphinx sits at an elevation, which places it very precisely at 15 degrees below this platform height.
>
> Finally, to further confirm the precise time this alignment occurs, our Sun must be in its exact location based on the required lineage line placement at precisely the same time. These 3 factors come into play only once and that date is, yes, December 21st, 2012.

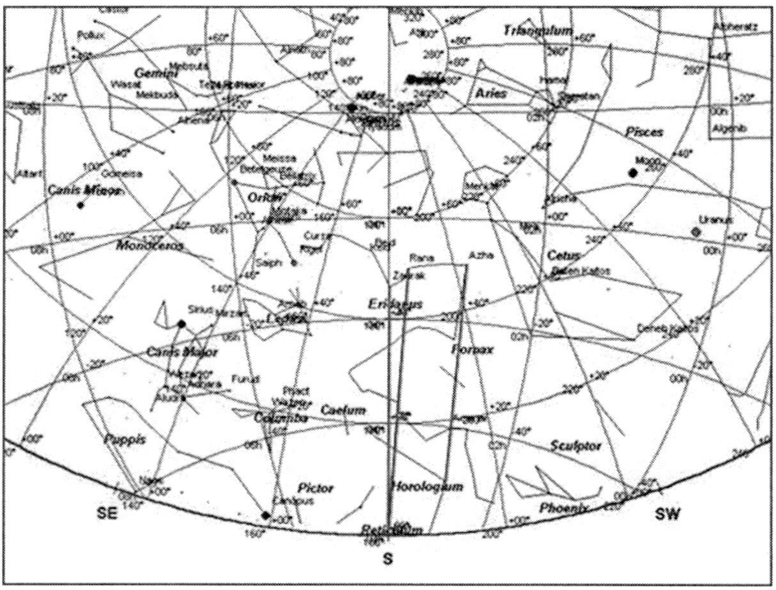

Illustration 3-1 Star map

Next, we checked the location due East of the Valley Temple of Khefre and its relationship as to its association equal in the skies at all time. Bearing in mind that although there is separation between the two Temples as they sit side by side, we find that also due East at 15 degrees below the horizon at the precise angle of deviation of 3.71425 degrees, it sits next to Rana. It is here that we discover the star Zaurak (meaning boat) is also a Star in the constellation of Eridanus.

Turning our attention now to the full plateau model as a further crosscheck reveals that an additional star in the Eridanus constellation aligns with our Sun at the precise time. By extending a line from a starting point from the apex of Khafre and through the apex of Menkaure pyramid (the East to West lineage set) this aligns precisely with the star Azha, meaning hatching place, (also of the Eridanus constellation). Further confirmation is obtained as we extend this same line in the opposite direction we find that this same line intersects precisely with our own Sun. While an extension of this line, in the opposite direction, aligns on past Azha and onto the Star Horologium.

Note that the time of this alignment is only a few hours prior to the precise end time alignment that occurs at precisely 22:18:13 hrs. (local Cairo time) on December 21st,2012. At which time a line drawn through from the apex of the pyramid of Khufu and extending down and out across Menkaure, will also then not only cross the Star Azha, but also we find that further extension of this line then aligns to the star Horologium (HOR) in the constellation of the same name. Once again our own Sun is in a direct alignment at this time.

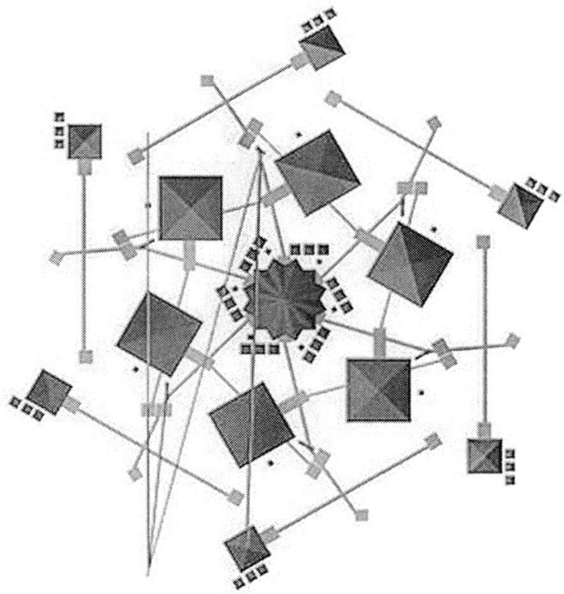

Illustration 3-2
Plateau model in relation to star map

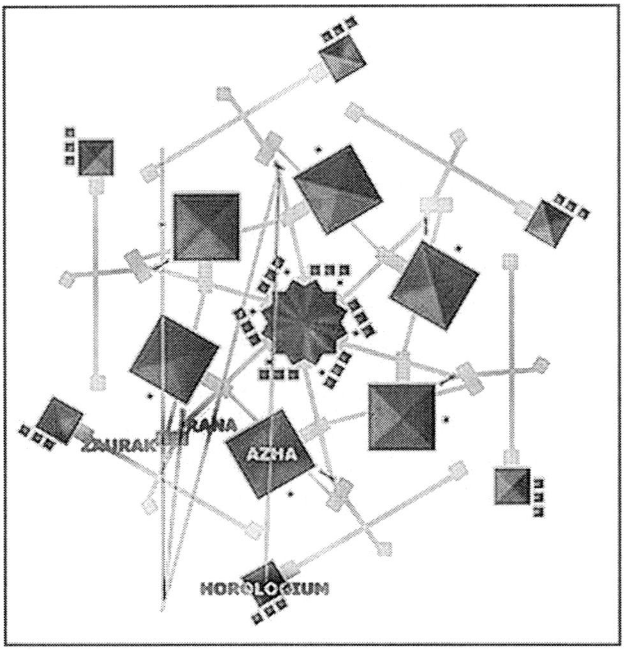

Illustration 3-3 Plateau model showing relationship of Horologium and stars of Eridanus

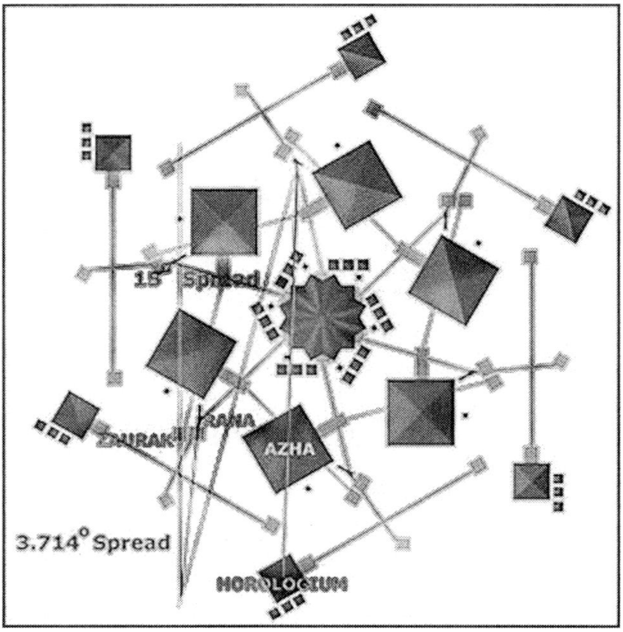

Illustration 3-4
Plateau model showing relationship of the stars of Eridanus and degrees spread

The reason for the apparent discrepancy in the alignment times is due to the fact that the artificial grid that was laid down by the Erideans aeons ago, will cease to be effective at this time. The time will reset at the time the full alignment is reached. This means that as the outer TWO grid lines will be aligning and as one begins to interact with the other the artificial grid (the one established by the Erideans) will no longer be able to interact or fit into that field as the two magnetic fields "0" points cross over one another.

Aside from what we have discussed above, we know of only one other reason for the Sphinx to be located on the plateau, and that is by the intersection point of specific extensions of lines up through the Sphinx giving the precise location of the now called Tomb of Osiris. Although we do not have the space available to elaborate upon the meaning behind this tomb at this time, it will be dealt with in extensive detail in Volume 3. Further note that the Sphinx and associated Temples lay outside of the Pi section of the model, which is explained in a later chapter in this book.

Further to this data we submit the following information: First, that during the time of the construction of the surface monuments, which occurred shortly after the planet was in the process of being made ready for habitation, forced labour was utilized to complete the erection of the majority of the monument markers that exist still to this day on the Plateau.

One may make an analogy here with the construction of the Tower of Babel, where;

> The whole world spoke the same language using the same words...
> They said to one another "let us mold bricks and harden them with fire..
> ....let us build ourselves a city and a tower with its top to the sky
> and so make a name for ourselves; otherwise we shall be scattered
> all over the earth. -Genesis 11 verses 1 - 4

Once the construction was completed, (in order to establish a marker) by slave labour under the supervision of the Erideans, the people were no longer required. The Erideans, being great mining engineers who carved out great underground complex's while terra forming and stabilizing planets, now left our planet leaving the city complex below the Giza Plateau vacant. Therefore, upon completion the race was scattered as explained in Genesis;

> Thus the Lord scattered them from those all over the earth and they
> stopped building the city. -Genesis 11.8

`It is a strong possibility that subsequent to this above event that the Anunnaki arrived on planet earth, being a result of the residual remains of their race from a light versus dark, or positive verses negative encounter leaving them with no home world. Thus leaving them trapped in this positive time frame, and seeing the chaos developing on earth, now took full advantage of the situation. From here it was purely a matter of time as this race began their influence, bloodline and agenda throughout the planet. Telepathically, and during a Remote Viewing session Edmund Meadows [3] was informed by the Erideans that they truly had used "Humans" as slave laborers. They regret that there was a great loss of life absorbed during this construction phase (they were referring to "human" life).

In front of the Sphinx an artificial Lake was created, where water from the river Nile was diverted in through a narrow channel, thus creating the lake itself. The symbolic concept behind this Lake and the adjacent river Nile is significant in that it helps explain the very reasons that the "Solar" boats (Barques) were buried below ground level. For the Lake signifies the empty expanse of space itself, and that the river Nile (being attached to that space in addition to the Plateau) is often associated with the constellation Eridanus. The significance of the Solarboats being buried where they are is quite simple, and is explained in book 3, where we explain the inter-dimensional cohabitation of this planet even to this date. But if one were to sail upon the ship across the great expanse of the lake, one would in fact find oneself at the river Nile. Meaning that if one were to traverse the distance across the open expanse of the waters (a universal open space expanse) one would be located in the Eridean's home constellation known as Eridanus.

> There arose out of the darkness a great shining egg and this
> was Re

At this juncture, it may be advantageous to remind the reader of the ancient Egyptian mythology surrounding the mythology of the creation myths, as it applies to our discussion.

Re or Ra, the Sun god, is said to derive from the creation of Atum, the first event, via Nu the primordial stagnant ocean, where Ra-Atum was the coming of the light and symbolized by the Phoenix. Could this in fact relate to the star RANA (Lt. Frog) in Eridanus? While Horus, son of Osiris and Isis, be related to the star Horologium (HOR) in the constellation of the same name. Horologium means clock, or time piece, does this mean that the December 21st 2012 is the re-setting of time itself? We believe this is so.

(Endnotes)

[1] There is the opinion by modern astronomers that:

1. Many of our comets may have originated in this constellation. A ring of dust has been observed around Epsilon Eridani which is "strikingly similar" to the inner comet-zone in the solar system. (Joint Astronomy Centre, Hawaii; Royal Observatory, Edinburgh, Scotland). It has not been clear as yet whether Epsilon Eridani has planets, but the newer images give a clue that there may be.
2. This constellation, along with it's neighbors can be seen all year round in the southern zones, south of 32 degrees N. Although one of the first 48 constellations drawn in Ptolomaic times, it was probably taken from the Ancient Egyptian traditions, as the Greeks would only see parts of Eridanus at certain times of the year. References by Eratoshenes, Germanicus, and Hyginus all refer to "Nilus" in this area of the sky.
3. RA —Right Ascension; hours measured along the celestial equator, starting from the March Equinox (intersection of the eliptic with the celestial equator).
   a.) Mintaka, the top star in Orion's Belt
   b.) Altair, in Aquila the Eagle
   c.) the pointer stars of the Big Dipper [Ursa Major]
   d.) Beta-Cassiopeiae, one of the outside stars of Cassiopeia.
   e.) Would it be meaningless to specify a Right Ascension for the north celestial pole?
4. ERIDU—Eridanus?
5. There are also modern mentions of the August Erianids being very old in reference to their orbits, and the radio-echo stream may indicate an extensive mass separation has occurred at one time in the ancient past. It has also been speculated that the Eridanus could have been the origins of some of the comets seen by ancient astronomers.
6. Except for the Alpha Centauri and Sirius systems, Epsilon Eridani is the star closest to us, at about 11 [eleven] light years distance. EE has been discovered to have a companion with a mass less that 0.05 that of the sun. It is separated from EE by about the same distance as Saturn is to our sun. This object may be a large planet or one of the smallest stars known.
7. This constellation, Eridanus, has been associated with many symbols in Ancient Egyptian religion/mythology. It resembles the 'serpent' or 'dragon' believed in very ancient Egyptian times, to have attempted to swallow the sun as told in the ancient and pre-historic Apep myth. APEP involved a serpent and was a "former discredited sun god". He had a companion, Mehen, associated with Set that was depicted in pre-dynastic and Old Kingdom art as a "serpent-headed man holding a spear, standing in the prow of the sun-barque". Later Mehen was known as the companion of Set/ Seth,[who was the pre-dynastic patron god of storms, desert winds, darkness (eclipse?), and who was worshipped as a benevolent god who helped 'save' the sun every day. Along with Apep and Mehen, these gods changed in either 'evil' non worshipped gods, or disappeared altogether

[2] Zecharia Sitchin Of the 12th Planet series fame, where he believes that the Sumerian and other ancient texts show familiarity with a complete solar system, that included beside the sun and moon ten (not nine0 planets. The tenth planet (or 12th planet member of the solar system) having an elliptical orbit that come around every 3000 years. The Sumerian mane was Niburu.

[3] Viking Remote Viewing

Chapter 4

THE GRIDS

In Volume 1 we discussed the relationship of the interior layout of the Pyramid Khufu and the three Girdle stones located in the ascending passageway, which were shown to be specifically and strategically placed. In fact, they represent their original intention, that of a doorway. Prior to the forced entry of Al-Mamun's men into the Pyramid, the action of which caused vibrational waves, the three girdle stones were originally holding back the three granite plugs. Once the girdle stones were vibrated they were "released" into their present resting place, that being adjacent to the base of the ascending passageway. This meaning that their full centre opening came directly into full alignment of the ascending shaft.

Each one of these three granite plugs represents an energy flow, each successively at a higher level. When the girdle stones were jarred (into their present position as described above) this action allowed all three granite plugs to freely move straight through the ascending passageway. Consequently, they struck and forced the smaller granite block that had been placed against the wall separating the ascending and descending passageways.

We will now explain the symbolism and meaning behind this extremely important and essential part of the interior plan. First, the three plugs represent quantitative energy and that once access is made the next space is made available and would then allow that energy to race forward into the next space. Secondly, the three girdle stones represent the outside perimeters that held back the full energy of the three plugs. It is not until the girdle stone itself is in alignment that this scenario would be possible, but once in alignment, this would itself create a doorway or "portal'. Finally, the effects of the granite block that was forced through by the impact of the entry. The effect of these three plugs on the granite block once they were in motion and the 'portal' created, was to force the granite block past its "placed" resting point. This being against the inside wall of the lower end of the ascending passage, thus forcing it through the wall separating the ascending passageway causing it to land in the descending passageway. These three plugs and their corresponding girdle stones represent the energy flow and consequent alignment of the three electro-magnetic energy grids that ultimately affect the granite block (represented as Mars) consequently appearing to move it to another level. In this instance, back out to interact and come to rest in the descending passageway. Therefore, thrusting it on and into a new pathway, or at least one that is set at a different associated angular alignment.

In volume 1, in the chapter titled "Interdimensional Physics", we explained the growth process in which we stated that the planet Mars, on its next shift will blow off its exterior core subsequently passing through the "river" or belt referred to as the asteroid belt. In turn it will pass through into the area where the planets Jupiter, Saturn, Uranus and Neptune presently reside and where this exists within an altered angle alignment plane to that of the inner planets of our Solar system. Pluto and all subsequent planets in the outer ring, of which we conclude there must be four, are once again at an angle relative to our sun, which in turn is at a further angle. In addition, we would state that all planets will be advancing, which will mean that as mass enters into the middle ring Neptune will be exiting into the outer ring.

We believe that each of the three sets of planets are set a 30-degree angle in total, however, Zero must always be factored into this calculation, therefore completing the 90-degree maximum. Each of these three sets of "inner" "middle" and "outer" planets are either viewed as positive or negative, the distinction being drawn by the polarity of the area of existence they are located within. We suggest that this is effectively controlled within the universal model by the three interlocking grids that govern this planet or planets in question.

This may be succinctly explained by the method of formation of matter and antimatter in space, where the waves travel through space in a spiralling motion, thus passing through alternate stages of positive and negative. Each spiral of 360 degrees creates a single pulse, where each pulse of physical matter creates a measurement, which is referred to as time. (Cathie 1983) These three areas described above and their associated interlocking grids that are set at 30-degree "maximum" angles are definitely proportionate to their directional flow. One will no doubt realize that what we are proposing is that our three lineages as explained through the decipherment of the plateau in Volume 1, are intersecting at their "maximum" 30-degree angles to each other. As the energy we are currently experiencing in this time and space, is both forward and under primary lineage it only leaves one to know when the maximum time is reached before change is affected. The remaining point of this concept is, at which points in time do the other two lineages affect their strongest influence on the time, space and matter continuum. Bearing the above in mind, we now demonstrate how the three grids are aligned and how through alignment will soon come into their maximum points of that alignment. In addition, how this process will open the door or portal, allowing the energy to enter from one level into the next level and so on. It will not be until December 21st 2012 that all three grids will have reached, opened then closed, or crossed over thus completing this alignment, there will however, be a series of alignments that will occur prior to this date. It is during these individual alignments between Grid 1 and 3, or 1 and 2, or 3 – 2 etc., the opportunity will arise for the entry of this energy, and into another space or time. In addition, this will be the time when "others" will also have the opportunity to enter "our" space-time line.

Bruce Cathie's diagram, which is shown to demonstrate the relationship of the polar grid squares, one may observe that to its outer corner, the E,I, M side bears a striking resemblance to the pyramid layout found on the Giza Plateau, where:

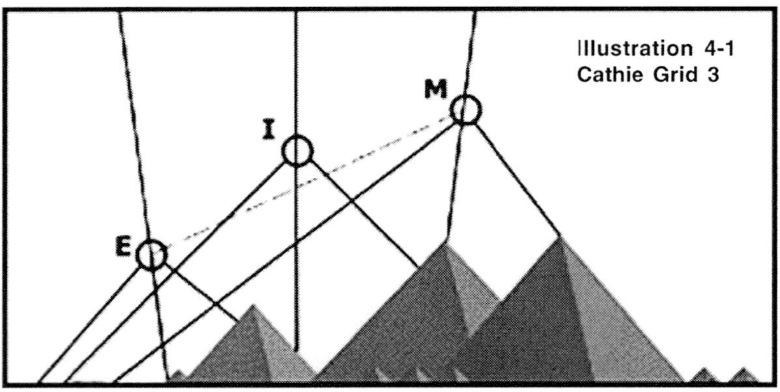

E - represents the Pyramid of Menkaure
I – represents the Pyramid of Khefre
M– represents the Pyramid of Khufu

Bear in mind that we identified these three very important segments that are, or could, be extracted from this precise size and placement on the Plateau. This is an extremely valid correlation, for if any alien race were to enter our space for the first time they could immediately identify the earth's grid lines. In addition, it would not only offer the layout of the earth's magnetic grid, but would also give the necessary data from which the time line can be calculated. This in turn would provide the same concept where the markings identified on the planet Mars would be pointed out to travellers entering that space. For additional information to the siting of the Sphinx in this layout, refer to chapter 2. At this stage we must point out that the diligent work of Bruce Cathie[1] in mapping out the magnetic grid lines surrounding this planet, certainly supports our findings. The reader may observe in the two diagrams that one does in fact support the other. The major difference in the two works is that one must incorporate the positive as well as the negative, both in function and alignment. These differences, and consequently the importance of the six grids (3 X 2) are explained in detail in Volume 3 of the Giza Genesis series.The only way to traverse the vast differences of space is to possess the means of manipulating, or altering, the very structure of space itself; altering the space-time geometric matrix, which to us provides the illusion of form and distance. The method of achieving this lies in the alteration of frequencies controlling the matter-antimatter cycles which govern our awareness, or perception, of position in the space-time structure. Time itself is a geometric, just as Einstein postulated; if time can be altered, then the whole universe is waiting for us to come and explore its nook and crannies.

DIAGRAM 3

Showing the relationship of grid polar squares A, B, and C. The polar squares are orientated in reciprocal positions around both the north and south geographic poles.

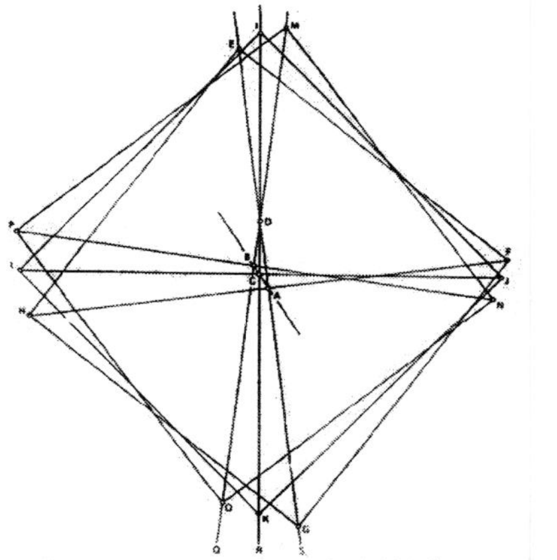

A: Grid pole "A": 1054.4 minutes of arc from geographic pole
B: Grid pole "B": 694.4 minutes of arc from geographic pole
C: Grid pole "C": 864 minutes of arc from geographic pole

D: North geographic pole
E, F, G, H: Polar square "A"
I, J, K, L: Polar square "C"
M, N, O, P: Polar square "B"

Q: Longitude 105 degrees west
R: Longitude 97.5 degrees west
S: Longitude 90 degrees west

Illustration 4-2
Cathie Grid 2

In the blink of an eye we could cross colossal distances: for distance is an illusion. The only thing that keeps places apart in space is time. If it were possible to move from one position to another in space, in an infinitely small amount of time, or "zero time", then both the positions would coexist, according to our awareness. By speeding up the geometric of time we will be able to bring distant places within close proximity. This is the secret of the UFO's – they travel by means of altering the spatial dimensions around them and repositioning in space-time.

-Cathie The Bridge to Affinity Page 22

> Einstein declared that physical matter was nothing more than a concentrated field of force. What we term a physical substance is in reality an intangible concentration of wave-forms. Different combinations of structural patterns of waves unite to form the myriad chemicals and elements which in turn react with one another to form physical substances. Different wave-forms of matter appear to us to be solid because we are constituted of similar wave-forms which resonate within a clearly defined range of frequencies – and which control the physical processes of our limited world.
>
> -Cathie The Bridge to Affinity Page 24

The "key" to understanding the weave of construct of the universe is as simple as understanding the meaning of the measurements as laid out in the Book of Genesis. These measurements are the crux to comprehending the interior of the pyramid of Khufu as it relates to the "inside" – "outside". The electro-magnetic grid lines governing the earth are derived in the same manner that all elements are controlled by and therefore must adhere to, in this universe in addition to parallel and other interlocking universes.

First, we establish a primary line as a set of five units and in our lineage we will use, as was intended, the pyramid Khufu as our centre pivot point calling it earth. The Pyramid of Khefre we will refer to as our sun, the sun being the birth mother of our Earth (in this time dimension). This is shown as being represented inside the interior of Khufu by the location of the Queens Chamber, located on the 25th course of the pyramid itself and where the entrance is on a direct lineage line. Next, we move up the lineage line precisely 25 courses and find ourselves at floor level of the Kings chamber, where this represents the birth mother of our sun, and where its angle from our sun, the Queens Chamber, is once again set at

26.42857 degrees. 26.42857 minus 1.42857 equals 25 degrees, once the factor of Zero has been factored into the equation.

As the Kings Chamber is not a direct parent (or "mother") it still requires a direct input into substantiating our Earth's existence. This is the very reason that it is referred to as the Kings Chamber. Because all measurements of Noah's Ark apply equally to all astral bodies, the negative, or opposite exists. This is also realized through alignment, and in the case of our earth we find the direct lineage line is also realized in this positive dimension as being split amongst two star constellations. The first of these stars being Azha in the constellation of Eridanus, while the second being a star that we will name Horologium in the constellation of Horologium. In addition, as these two stars are not directly related to one another it is only through their alignment that at a precise time they fall into our line of direct influence. This then becomes our negative half of our lineage line as it is in direct opposition of the sun side of our lineage.

The five tiered ceiling above the Kings Chamber indicates that this sits on top of our five tiered model, while the five tiered niche located in the Queens Chamber opening, in a downward direction, signifies the full alignment of the five tiers. This then allows for the birth channel to open. Our main lineage line when viewed from the side, as the illustration demonstrates, is similar to a five-tiered step. The smaller three "queens" pyramids of Khufu and Menkaure are those that exist on either side of our "mirrored" single linear line.

The subject of the connection of these pyramids and their relationship in establishing the positive and negative points of alignment, and their reference to the multi dimensional layout of the grids, will be discussed in some detail in Volume 3 of the Giza Genesis series. However, at this juncture we will concentrate on the layout of the model as a whole in two-dimensional form, utilizing 3-dimensional imagery wherever possible. Again, in Volume 3, we will discuss the subject of the positive and negative rotational spin, but at this point we are more concerned with the stationary fixed grid alignment, as it will arrive into maximum alignment on December 21st 2012. At that time it will have come to its full cycle thus crossing over and sealing.

Again, the numbers to look out for are 50 – 51.42857, 25 – 26.42857 and of course the maximum 30 degrees, this then demonstrates the formation of the grid in all three lineages in maximum alignment.

> Make yourself an ark of gopher wood, put various compartments in it, and cover it inside and out with pitch.

This is how you shall build it: the length of the ark shall be three hundred cubits, its width fifty cubits, and its height thirty cubits.

Genesis 6, 14-15 Preparation of the Flood

In our thesis we associate the flood, or river, with a surge of energy produced when the grid lines cross over.

All the foundations of the great abyss burst forth.
And the floodgates of the sky were opened.

Genesis 7-11

It is recorded in Genesis that the rains persisted on earth for 40 days and 40 nights, a time we believe when the full grid alignment will have been "centered" and occurred. It is interesting to note that a similar flood was recorded and preserved in the eleventh tablet of the Gilgamesh epic. The floor of the Kings chamber is positioned precisely with the 50th tier level of the Great pyramid, while the sarcophagus is sited at 1.42857 degrees above that point which the calculations showed in Volume 1. The angles of the descending and descending passageways when calculated from the outer edge of the casing stone verses the interior setting, is set at 26.42857 degrees. The small block placed at the point separated originally by a wall, where the ascending and descending passageways intersect, represents the planet Mars.

The maximum angle from the outside edge of the outer casing stone to the apex of the Kings chamber is set at precisely 30 degrees, while the apex of the Queens chamber is set at precisely (when measured from the same location) 15 degrees. Thus making the maximum height 30 degrees.

Therefore, taking all the above into account, we may interpret this to mean that, according to our model, each level of this expanding stratified complex weave is just as consistently governed at each level, as the constant set in numbers indicate. Thus from our Zero point (Earth's interior crust) the alignment of the electro-magnetic grid to the centre of the core is set at the same angle as the six closest aligned stars, that being 26.42857 degrees. While in turn, each of these six stars are themselves aligned to the stars which justify their existence in time and space, which are equally set at 26.42857 degrees from its core.

One could ask does this 26.42857 degree angle (centre point to outside) which when minus "0" or 1.42857 equals 25, or half the 50 which is mentioned as the width of the Ark? If this is indeed true, then the 300 (length) must represent the overall direct lineage. This then could mean that the sun and Azha (in the constellation of Eridanus) each at 26.42857 degrees, one being positive the other negative, less the values of "0" which equals 25 plus 25 which equals 50. In turn added to their respective sun's origins angle of 51.42857 less "0" (1.42857) that equals 50 x 2 = 150 X 2 (300) the length of the Ark.

As we pointed out in volume 1, once the star code had been correctly interpreted we found that Noah's Ark actually demonstrated to be the Pyramid of Khufu. For once it was placed into the model at the correct angle, all shafts aligned to the specific stars.

What we have been able to decipher from the inner moving parts of the Pyramid of Khufu follows below in a succinct format. Once the trigger is set in motion, possibly by the upcoming electro magnetic shift, the reaction would have produced sufficient vibrational motion, which in turn would set the string of events into action. Unfortunately, no one has accounted for the chance occurrence of a forced entry into the pyramid that would trigger, or jar, part of the mechanism to prematurely activate. As we have discussed the granite block previously in this chapter, in addition to the girdle stones and granite plugs, at this juncture we will deal with what type of reaction would have been produced behind the original location of the upper granite plug.

In chapter 1 we mentioned that behind the last plug Al Mamun's men encountered three additional blocks, however, these three blocks were made out of the softer limestone. These limestone blocks were then broken down into pieces as Al Mamun's men moved on forward into the Pyramid. Once past these three blocks, they then found themselves entering into the Grand Gallery, whereupon they encountered the present opening leading towards the Queens chamber passageway. This area in front of the Queens chamber, as we explained in Volume 1, is known as the first veil. We further explained that the second veil remains closed to this date.

Had nothing previously been released prematurely, once the magnetic shift were to occur, the following events would have been triggered in the following manner; First, the second veil would be dropped which in turn would allow the release of a block which at the time of writing, is being held back from entering into the passageway. This block represents the birth of the new planet that has yet to exit from within our own sun, which will occur once the natural magnetic shift were to occur.

This new block would represent our yet to be realized "birth event", which would then land in its location directly behind the three limestone blocks, where the impact of such an event, would in turn force the blocks forward down the ascending passageway. They will continue forward until impacted against the last granite plug where the chain reaction would continue on until the granite block was moved out into the descending passageway.

What actually lies behind that limestone block, being held back by the second veil, will be disclosed in Volume 3.

In order to relocate into a specific star's real time line, one merely has to be fully aware of the local grid alignments that directly control both the present time line one is located, in addition to the point where one wishes to enter. Once having identified the respective grids, it would then simply be a matter of advancing to any of the intersection points of these two governing grids. Once this is accomplished one then sideslips into the other parallel time line. However, there is a possibility of a miscalculation of the timeline by failing to factor in the variance of the positive/negative fields.

As this is the basis behind the driving force behind the very battle we are about to find ourselves entering, we must come to understand the mechanics of the hows' and whys of this scenario. The subject of the positive and negative sidelines is dealt with in detail in volume 3, however, in this present volume, we will attempt to show how "some" of the observed phenomena may be translated.

Once one is able to establish the exactness of each individual piece of the puzzle, then, and only then, may the pieces begin to fit and arrange themselves to help complete the overall layout of this very "cosmic puzzle".

(Endnotes)

[1] Cathie, Bruce The Bridge to Infinity Harmonic 371244 Brookfield Press New Zealand 1983

Chapter 5

THE LINE HAS BEEN DRAWN

The running thread throughout all our previous research has consistently been the interpretation and connection of the Giza Pyramid complex and the association of the books of Genesis and Exodus. We have offered an explanation that in connection with the available evidence, both set in stone and in text form a set of data was handed down for the benefit of mankind. How the many groups involved in related research have interpreted the data, is as varied as the number of chapters in both Genesis and Exodus itself.

However, these sets of data, correctly evaluated and interpreted, will assist mankind towards an understanding as to not only what scenario is currently unfolding, but to future events, visa vis the date of December 21st 2012.

We have offered a plethora of data for you to analyze and check out, including planetary and stellar alignments as related to the very siting of the whole of the Giza Pyramid complex, in addition to the historical and 'mythological' information.

In our chapter INTERDIMENSIONAL PHYSICS in Volume 1, we demonstrated the importance of the December 21st 2012 date offering an explanation towards the relation to the alignment of the heavens as laid out on earth at the Giza complex. In addition, we also emphasized the importance of the May 5th planetary and stellar alignment, also calculated from the Giza complex and mirrored in the heavens above. This theme was continued in the chapter ON THE EIGHTH DAY, Volume 1 where we went to great lengths to sum up the current situation with the available evidence via the many links on the Internet..

In one way, this is an attempt to collate the many themes we have been discussing to present into the public domain this essential offering towards an understanding as to our choice of free will.

When we originally wrote this section Brookhaven National Laboratories, home of the Relativistic ion collider situated on Long Island New York, were preparing to instigate an experiment that will in effect reproduce, or so they believe, the so called 'BIG BANG' origins of our universe. If you recall we explained in the chapter ON THE EIGHTH DAY how the blue grid was activated around planet earth, and the possibility of Brookhaven firing up the collider to punch a hole in that grid. If this attempt was successful, then the particular race waiting to arrive on our planet earth, will make their presence known and instigate the fall of Terra.

This then, may be the time that the 'deal' struck some time ago (SEE On the Eighth Day) by a 'elite' few, not the people themselves, will be 'called in'. If this scenario was allowed to go through, the 'war in the heavens' will enter our part of the Solar System, and we will be drawn into the conflict.

The whole point of this 'deal' was in fact a great deception in itself, an excuse to set in place the technology to allow the eventual take over of Terra, thus enslaving her people and preventing our ascension, on December 21st 2012.

Before we submit a detailed explanation as to how events under the Giza Plateau on May 5th may have given us the opportunity to change this scenario, we place the last paragraph from our previous work as a stark reminder as to what may/may not occur. The important point is that YOU HAVE that option - to exercise FREE WILL.

> "By being prevented from moving to the next level, we remain where we are, and on the course we are headed. This would result in the near extinction of the people of Terra, leaving the planet and it's resources to those who would desire this piece of real estate. Those that remain after this take over, would become drones for the new occupants. We have made the choice not to allow this!"

To continue:

We now offer an explanation and an understanding as to how all this relates to the 'Giza-Genesis' connection, we submit the following background. In 1998, rumors began to circulate, then escalate on the Internet concerning the anomalous sudden appearance of a vast number of white, unmarked vehicles witnessed parked in storage sites around the USA. Later, photographs of these vehicles began to appear. As interest grew and investigations commenced into this phenomena, an influx of similar white vehicles were observed with

UN (United Nations) markings on the vehicles, even been photographed being loaded onto rail cars for dispersal.

It was even reported and photographed, that in many cases shackles were seen attached to these vehicles, and some observed as 'prisoner carriers'. For photographic evidence of this and the fact that a number of 'prisoner of war' camps were being set up around the States please link to news watch magazine:

Events began to get confused when it was reported that school buses were also being converted into UN troop carriers. This scenario began to get even more confusing when it was discovered that Russian soldiers were on US soil, said to be taking part in some 'joint operations'. Not only were photographs taken of Russian vehicles being transported on a Texas highway, but the discovery of a 'secret military base' was also unveiled.

"Mississippi civilians have discovered a secret military base, manned by civilian and U.S. and foreign military personnel, deep within the DeSoto National Forest in the south-eastern part of the state." An extract from the report by Mike Blair is included here.

Secret Base Is Discovered

If there's nothing nefarious about the government's intentions, why hide military activities in a preserve for endangered species?

Exclusive To The Spotlight By Mike Blair

Mississippi civilians have discovered a secret military base, manned by civilian and U.S. and foreign military personnel, deep within the DeSoto National Forest in the south eastern part of the state.
After discovery of the secret site, the Mississippian patriots, dressed in camouflaged clothing and equipped with cameras and videotaping equipment, were able to penetrate the outer perimeters of the base and have provided The SPOTLIGHT with numerous photographs and about 45 minutes of video tape.
The base, located within a restricted area of the national forest, posted with signs designating it as an "endangered species" habitat, covers a considerable area. Abundant about the base were Russian T-72 main battle tanks, BTR-60 Russian armored personnel carriers, a Scud surface-to-surface missile battery consisting of several individual units and missiles, at least one Russian SA-6 Gainful surface-to-air (SAM) missile launcher with its three missiles elevated and in launch position and a Russian SA-8 Gecko SAM carrier with

its four missiles in place.

Armed Civilians

The camp at the time seemed to be guarded primarily by "civilian types," who carried side arms in holsters.

A retired U.S. Special Operations officer told The SPOTLIGHT that the operation sounded to him like some type of "top secret spook [CIA]operation." All of the Russian equipment, as described above, was photographed with both conventional and video cameras by the Mississippians, who have dubbed themselves the "Confederate Marines." A Pentagon spokesman told The SPOTLIGHT that he had no information regarding any type of military operation in Mississippi's DeSoto National Forest. The obviously top secret operation was discovered by the Mississippians August 15 when they encountered a long military convoy on a remote highway in the DeSoto National Forest area. The vehicles were not only on the road but dispersed in the woods on both sides. The civilians tried to engage in conversation with troops in the convoy who were garbed in "foreign, probably Russian" military camouflaged uniforms, but found they could not speak English.

Taken from http://www.webbindustries.com/spotlight/f_mb_art002.html

For a further insight, read the interview with a German soldier, where it was alleged that German troops were stationed at Holloman Air Force Base, Alamagordo, New Mexico

David went to investigate the rumor that German troops are supposedly stationed at Holloman Air Force Base in Alamagordo, New Mexico. When he arrived, he met a German Soldier (pic above) who volunteered to be interviewed. He cheerfully answered all David's questions until David asked him straight out, "Are you being trained to occupy America's streets when Martial Law is declared?" Immediately afterward, the soldier raised his hands up, turned away, and left without speaking another word. Did David ask the right question?

http://www.newswatchmagazine.org/pics/holloman.htm

where the question was put;

"Are you being trained to occupy America's streets when Martial Law is declared?"

Naturally, curiosity and concern was rife, what was the purpose of this scenario unfolding? Later in the article we offer an explanation as to why these events began unfolding.

Let us now examine a further related event in the summer of 1999, to be precise at White Mountain, Phoenix an event that was termed "The White Mountain Evacuation".

It was reported in the July that an evacuation procedure notice was given in the White Mountain area, where approximately 1500 residents received a mailing from the US Department of Interior, giving them notice that they were subject to a possible 20-minute notice of evacuation. However, much of the information coming out was garbled and confused.

White Mountain is also said to be the location of a large underground military facility.

Independently, this event may be discounted as any reliable 'evidence', however, coupled with a plethora of additional information, one must seriously consider the implications.

It was interesting to note for example, that reports were coming out that the 101st Airborne had been on alert for some weeks previous to this evacuation order, and that Cheyenne Mountain (Home to NORAD North American Aerospace Defense Command) as on Defcon 3 status. The reference link for the majority of this data was located on various Internet sites, however, many of the links have now been pulled off the Internet.

Having taken on board the background scenario that has been taking place, we now move to a sensitive issue and one that has been increasingly debated of late, that of 'Chemtrails'. Commonly called 'Sky painting', this anomalous and increasing activity is causing serious concern amongst many citizens. A good overview is placed below:

Chemtrails - What You Need To Know - New Summary And Medical Data

From Holly Deyo hollyd@netconnect.com.au

There are few issues today that have sparked the controversy inspired by chemtrails. It's a near-global event. Three main theories are circulating to account for the spraying:

1. weather modification 2. mass depopulation 3. mass inoculation

Frankly, Stan and I didn't know where to stand on this issue several years ago as there was too little factual information though we leaned toward the inoculation theory. This was during the time when our armed forces were being inoculated for anthrax and a holy furor ensued. We had, and still do, have a hard time believing our government is out to exterminate the citizens.

At least part of the spraying seems to be for weather modification and/or neutralizing global warming effects according to recent interviews with Will Thomas. He further purports that this weather mod is being conducted due to pressure from insurance companies who see financial ruin on their horizon with continued high disaster-related claims. The catch? These weather modification tactics are further depleting the ozone hole and enhancing global warming. It is a short-term, "band-aid" answer with even worse consequences.

It appears there are two types of spraying conducted that would help account for several perceived scenarios. For those to whom this information is relatively new, here are some basic _generalizations_.

1. Chemtrails, as opposed to contrails, take three or more hours to dissipate. Regular contrails disappear in about 20 minutes.

2. Chemtrails tend to look flat.

3. Another feature different from ordinary contrails is that mature chemtrails display a vertical curtain of fall-off.

4. Intense chemtrail spraying is done in tic-tac-toe, "X" and grids patterns.

5. Spraying is generally done by these planes: KC-10A, KC-135 (707) and C-130 for the high altitude spraying and C-130 turboprop aircraft for the low-level spraying. The KC-10A is a three-engine aircraft and would normally leave three parallel contrails in its wake. The KC-135 is a smaller, four-engine aircraft that would typically display four parallel contrail lines. When used during chemtrail missions; however, these aircraft lay down only a single, thick, billowy chemtrail while simultaneously displaying no normal contrails. There are other planes which have been used in the spraying, but these are the main ones.

6. Two types of spraying are done: high altitude meaning 20,000 feet and above and low altitude spraying each with seemingly different purposes.

7. Low level spraying contains molds and fungi and produces varying symptoms including: * headaches worse than migraines * heartburn * heart conditions * heart arrest, on the _rare_ occasion * flu-like symptoms * extreme fatigue * diarrhea * pain and swelling of muscles and joints * dizziness * nausea * stiff neck * nagging sore throat * general malaise

8. Symptoms last up to three months and beyond, not typical of flu.

9. Symptoms are not readily treatable by known antibiotics.

10. Chemtrail missions tend to be suspended when the ground wind speed reaches or exceeds 20 miles per hour.

11. Tends to kill the elderly, the very young, and those with poor immune systems.

12. After spraying, gelatinous material is seen from low-flying, multi-engine aircraft. Cobweb-like material from higher-altitude spraying has been analyzed in university government-licensed labs with some of the pathogens identified.

13. Countries reporting spraying include Australia, Canada, England, France, Germany, Holland, Ireland, Italy, Mexico, New Zealand, Scotland, Spain and the United States. NOTE: This spray list does not include China,

Russia nor the Middle East. The US has reported spraying in virtually every state and virtually every major US city including surrounding suburbs.

WHAT'S IN CHEMTRAILS?

Part of the substances in the chemtrails have been identified: a cocktail of JP8+100 jet fuel, laced with Ethylene Dibromide (EDB). This chemical pesticide was banned in 1983 by the EPA as a definite carcinogen and chemical toxin. Exposure to this type spraying can include these symptoms: * respiratory tract problems * severe infections to the throat and sinuses * swelling of the lymph glands * coughing fits * shortness of breath * sinus headaches * general respiratory failure * damage to the heart and liver * exposure to EDB makes people more susceptible to other biological agents due to severe lung irritation

For lack of a better term, "brown goo" has been used to describe the material found on exterior walls of buildings and windshields of automobiles. It's almost impossible to remove with soap and water, and has proved to be highly toxic to anyone coming in contact with it. Don't touch it, handle it, or attempt to move it by yourself.

"Numerous red and white blood cells, and unidentified cell types have been found within the sub-micron fiber sample previously presented and submitted on Jan. 20, 2000 to Carol M. Browner, Administrator of the United States Environmental Protection Agency. The cells appear to be of a freeze-dried or desiccated nature in their original form within the microscopic fibers." http://www.carnicom.com/bio1.htm

"Thomas has also reported that spraying samples have been analyzed and have revealed that many deadly and toxic pathogens have been found including Mycoplasma Fermetens Incognitus (the SAME bioengineered pathogen that Dr. Garth Nicholson discovered in about 45% of the veterans who came down with Gulf War Illness). Thomas found that Mycoplasma, however, was only ONE pathogen among a group of highly toxic biohazard substances analyzed from the chemtrail residues."

http://www.educate-yourself.org/ctindex.html

Basically, the scenario is one of large aeroplanes trailing anomalous 'jet' vapors, which seem to spread thickly over our skies, and said to contain some peculiar cocktail of chemicals and bio agents of unknown origin. The result of this 'spraying'; activity (in some circles known as 'Operation Cloverleaf') is that many people suffer from a variety of symptoms, ranging from irritating coughs and dry throats to more severe debilitating effects.

It is reported that this atrocious method will not only continue, but expand into a full-blown 'bio attack' scenario, leaving citizens completely incapacitated! Why? And what has this to do with the articles on the Giza Plateau complex?

Let us now examine the profound findings that have come out of our research at Giza to place forward an explanation towards an understanding, and a preparedness, for an imminent (and we mean imminent) decision incorporating one's Free Will.

We described in On the Eighth day how the Global grid activation took place, originating from the focal point at Giza, and how this began the trail for the opportunity of Free will choice. The decision one has to make will now be explained, as will the current situation regarding the agenda of the remaining covert 'government' concerning all who live on this planet.

THE EXODUS CONNECTION:

The version of the Bible
Moses

Chapter 4: 2
The LORD said to him, "What is that in your hand?" He said, "A rod."

4:3
and he said, "Cast it on the ground." So he cast it on the ground, and it became a serpent; and Moses fled from it.

4:4
But the LORD said to Moses, "Put out your hand, and take it by the tail" —so he put out his hand and caught it, and it became a rod in his hand.

Chapter 7:
Moses and Aaron visits the Pharaoh demanding that the people be set free, he throws down his staff and it turns into a serpent, in reply the magicians of the Pharaoh carry out a similar feat, but the snakes of Moses consumes the snakes of the Pharaoh

7:10
So Moses and Aaron went to Pharaoh and did as the LORD commanded; Aaron cast down his rod before Pharaoh and his servants, and it became a serpent.

7:11
Then Pharaoh summoned the wise men and the sorcerers; and they also, the magicians of Egypt, did the same by their secret arts.

7:12
For every man cast down his rod, and they became serpents. But Aaron's rod swallowed up their rods

Phase 1 has begun, the Alien Red grid was placed up around planet earth, but then on May 5th, the 'Blue grid' was activated, thus overriding, or consuming, the red grid.

Phase 2
Pharaoh, not wanting to give up his slaves (Jews, the light Force Souls), orders the first-born (son's) killed

Chapter 1:22
Then Pharaoh commanded all his people, "Every son that is born to the Hebrews you shall cast into the Nile, but you shall let every daughter live."

Moses responded with the plagues that directly affect only those connected to the Pharaoh.

The use of the word 'Jews' is that of the 'light Souls' of the inhabitants of Jerusalem.

Revelation states that once all things have come to pass, Jerusalem will once again enter into the thousand-year reign. Meaning that all three zeros will align and from that time will once again begin.

From current events presently being enacted, we know that the New World Order (N.W.O) under the control of Esau, have plans to chemically attack Joe public. In doing so they hope to eradicate the first born, hence the reason in most cases for these Chemtrails affecting women more heavily, such as recent cases seem to be proving out.

However, this scenario will backfire on them. Being mainly of the reptilian descent they live but one physical life however extended that may be, and thus require tapping into our life force in order to sustain their existence.

The Ultimate Finale.

The Parting of the Red Sea

Exodus 14: 16

Lift up your rod, and stretch out your hand over the sea and divide it, that the people of Israel may go on dry ground through the sea.

The red sea is equated to the contrived alien red grid, which is parted and the Jews (light force souls) escape by passing through the opening, while the soldiers of the Pharaoh are trapped, thus confined to the banks of the red sea, or the red grid. Which in this case will occur on June 22nd 2000!

This is the scenario now being enacted out, unless they change their ways, they will not be able to follow through the parting of the red grid. Hence, all the decisions made in the past will flood back to haunt them in turn influencing their chances of a non-bias decision.

When this race visited this planet aeons ago, they interfered with its natural rhythms they manipulated humankind by genetic experimentation, using man as slaves to carry out their 'dirty work'. Thus, we have not been able to develop our psychic abilities that enable us to see the real truth of who is manipulating and deceiving the population (Matrix).

Fortunately, more and more people are awakening to this real truth; therefore the human race will soon be able to choose their destiny in full knowledge and free will.

For an additional take on the current situation, it may be advantageous to examine what the book of Revelation has to say and to briefly analyse the relevant sections in relation to our work.

> "And I heard another voice from heaven, saying, ***Come out of her, my people,*** that ye be not partakers of her sins, and that ye receive not of her plagues." (Rev. 18:4)

The white horse and Faithful and True stand ready to make their entrance, to judge and to make war. His armies follow and wait in the wings.

Those who hold the Covenant with God will soon be called to make the decision of whether to leave this planet or to stay. If they stay, they will be called to fight the battle of Armageddon.

> "And the woman was arrayed in purple and scarlet color, and decked with gold and precious stones and pearls, having a golden cup in her hand full of abominations and filthiness of her fornication. And upon her forehead was a name written, MYSTERY, BABYLON THE GREAT, THE MOTHER OF HARLOTS AND ABOMINATIONS OF THE EARTH." (Rev. 17:4-5)

Does this not sound like our Earth at present...arrayed in two colors, red and purple, in the form of grids?

> "...and they that dwell on the earth shall wonder, whose names were not written in the book of life from the foundation of the world, when they behold the beast that was, and is not, and yet is." (Rev. 17:8)

Those who were not and are not of the Covenant will not understand the call. For all will be called, but not all will hear.

> "These shall make war with the Lamb, and the Lamb shall overcome them: for he is Lord of lords, and King of kings: and they that are with him are called, and chosen, and faithful." (Rev. 17:14)

Those called may make the decision to leave Earth before the battle. A vortex will be created in the red grid, and the ships will gather them up, off this planet.

> "But lift thou up thy rod, and stretch out thine hand over the sea, and divide it: and the children of Israel shall go on dry ground through the midst of the sea." (Ex. 14:16)

> "And the fruits that thy soul lusted after are departed from thee, and all things which were dainty and goodly are departed from thee, and thou shalt *find them no more at all.* (Rev. 18: 14)

> "Rejoice over her, thou heaven, and ye holy apostles and prophets; for God hath avenged you on her." (Ref. 18:20)

There were two comets that came into view, the first being, 76P West-Kahoutek-Ikemura, that headed towards Mars on June 5th 2000, the second, C/1999 S4, was at perihelion in July 2000.

> "And a mighty angel took up **a stone like a great millstone**, and cast it into the sea, saying, Thus with violence shall that great city Babylon be thrown down, and shall be **found no more at all**." (Rev 18:21)

Many visionaries are now receiving visions of one or two comets that are destructive to areas of Earth.

> "And **the remnant were slain** with the sword of him that sat upon the horse, which sword proceeded out of his mouth; and all the fowls were filled with their flesh." (Rev. 19:21)

> "And the light of a candle shall shine no more at all in thee; and the voice of the bridegroom and of the bride shall be heard no more at all in thee: *for thy merchants were the great men of this earth; for by thy sorceries were all nations deceived.*" (Rev. 18:23)

The merchants, who made the deal with those who want this earth, have deceived many. They are but few, but many have followed them.

The people of the Covenant, who stay to fight the battle, will not be left here afterwards. For the ships will come and after the battle, those who died in body, and those who survived, will be taken up to the ships also.

And where are these ships going?

There is another planet made ready. The Earth will ascend, without those who made the decision to follow the beast of the merchants. This ascended planet is not the physical Earth we inhabit today, for that will be destroyed.

> "And I saw a new heaven and a new earth: for the first heaven and the first earth were passed away; and there was no more sea. And I, John, saw the holy city, new Jerusalem, coming *down from God out of heaven*, prepared as a bride adorned for her husband." (Rev. 21:1-2)

And how is a bride adorned for her husband? Cleansed and without physical ornament.

The ships come closer each day. Joseph stands ready to open the books to see what is written on whom are the people of the Covenant.

Very soon, the hearts of man will be weighed. Very soon, the people will have to make their choices. Those people of the Covenant who do not choose at all, will stay, and will be taken after the battle.

> "And I, behold, I will harden the hearts of Egyptians, and they shall follow them:..." (Ex. 14:17)

> "I have seen the aftermath of what comes. Everything flattened in a field of mud and blood, deep enough to reach my knees. By looking into the eyes of the bodies lying there, it is obvious who is to be

taken. They are lifted to the ships above. The ones who are to be left, are dead, both in body and in soul. There are no survivors except those that are taken to the ships." (J.K)

There is not much time left before the battle begins. Joseph has his tools ready, and he has called the ships to come. They are on their way and wait outside both grids at the present time. They will very soon be visible below the blue/purple grid, and above the red grid. At this point, after the books are read by Joseph, he will raise his rod and put out his hand and a vortex in the red grid will occur, allowing the ships to enter and take the people of the Covenant who choose to go.

At this time, Jacob's ladder will descend to bring the angels to earth to fight the battle of Armageddon. The white horse carrying Him who is called Faithful and True will arrive to lead the armies of those who decide to fight. And He will take his people with him, once the battle is over.

Do you see? Do you hear the call? What will be your choice?

For those that break the Covenant will die an Everlasting Death. And those who uphold the Covenant will arise to Everlasting Light.

Stand true and ready-He comes.

Whereas many people assumed the ascension would be one of mainly a spiritual nature (which of course in part it is) the actual 'exit' from this planet when the time comes is very much a physical one. That is the entry through the grid of many large ships that will enable those prepared and willing to 'ascend' to their destined pre-arranged especially prepared location.

Basically, it is understood that this planetary ascension would be by a method known as "disintegration", as this section from Julie King elaborates:

Travel by thought, with no space and no time involved, does not require movement, as we know it. What is required is the matching of vibration of the thought to the vibration of the grid.

To travel to any planet as the vibration of that planet, and the resulting planet form, one must go through a series of portals in the grid for that particular planet. Though the travel is instantaneous, the particles do travel through a series of mathematical portals.

To travel to Earth becoming Earth's vibration with Earth physicality, the portal is the triangle created by the stars Alpheratz (Andromeda), Arcturus (Bootes), and Antares (Scorpio). As such, all souls coming to Earth at human birth, travel through this gateway. These matrix points create the mathematical sequence portal for this planet, and the planet sits at a further particular grid point.

As a human being, to travel via the grid has the following sequence:

1. The thought occurs to travel to a point of space or time.
2. This thought is a mathematical energy.
3. The body disintegrates, into small pieces. These pieces are pure energy, and not visible.
4. The thoughts of the being disintegrate into energy. The human can no longer create a thought.
5. The thoughts re-integrate at the proposed thought grid point. The human can now think again.
6. The body re-integrates at the proposed thought grid point as a physical entity.

If someone were observing this event from the original grid point, the person travelling would just seem to disappear!!

Further, once the body has left planet earth, there is no longer the restriction that one has become used to.

The person we refer to as Joseph, who was awakened from stasis from below the Giza Plateau on May 5th, apparently has gathered his troops in readiness for his confrontation with one Esau.

Esau is now aware of his awakened presence and the call has gone out to Esau.

"Come...let us battle...let us carry out the Plan." "Come, Esau, come now....you will have your Earth....but

You Will Let MY People Go"

THE BATTLE LINES ARE DRAWN

THE ASCENSION:

The Oxford Dictionary describes this event as thus:

- The act or instance of ascending - to move upwards
- **The ascent of man, an upward movement or rise and advance or progress**
- Biblical - the ascent of Christ into Heaven on the fortieth day after Resurrection.

On June 22nd 2000, at 00.59 a stellar alignment coincided with the physical layout of the Giza Pyramid complex. This alignment will be the marker to instigate the initial phase for what some term 'The ascension'. This is the time as it will be in CAIRO, EGYPT.

We discussed in recent articles the May 5th alignment and in particular how access to the chamber below the Giza Plateau was carried out via the layout of the Bethlehem causeway, as reflected in the Betelgeuse alignment in the Eighth Day.

We will now demonstrate how the stellar alignment reflected in the layout of the Giza complex on **June 22nd** 2001, relative time, will instigate the preparatory phase of the ascension of humankind.

In our chapter Door No 2 Cayce's Revenge in Volume 1 we mentioned the location of a number of tombs below the Plateau, including that of Osiris and Joseph. If you recall, some time ago it was reported that the 'representational tomb' of Osiris had been discovered approximately 100 feet below the causeway of the Pyramid of Khefre. This was covered as a "LIVE" event via FOX TV.

The stone sarcophagus was discovered under water and cornered on four sides by pillars.

In the alignment on June 22nd the representative four pillars, corner to corner line up the stars to the smaller pyramid Menkaure.

To analyse this alignment one must first bear in mind the following procedures. The present position of Menkaure must be 'moved' 120 degrees, in the opposite direction of the Bethel placement (Bethlehem causeway as discussed in the chapter Eighth Day, and drawing a line across the apex of Khufu (Cheops). This will make the point of intersection on the tomb of Osiris. In addition, at 90 degrees the two shafts that exit from the Kings chamber of the Pyramid of Khufu are set at 45degree angle, the apex of the pyramid being the zero point.

Now let us examine this alignment in more detail. The two constellations that we are concerned with are **CYGNUS** and **DELPHINUS**.

Cygnus, the Swan, is one of the two birds hunted by Hercules, Aqua the Eagle being the other, and where both birds escaped!

The assumption is that these two birds, together with the vulture, form the constellation **LYRA**, the Stymphalian birds, and one of the tasks of Hercules.

Interestingly, Cygnus the Swan is also identified with Cycnus the son of Mars, who transformed the river Padus and transported to the sky.

The main stars of CYGNUS
Sador; Sadir. Gamma HR 7796 HD 194093
Gienah. Epsilon HR 7949 HD 197989
Azelfafage Pi 1 HR 8301 HD 206672
Al Rukbah al Dajajah. Omega 2 HR 7851 HD 195774

Our second constellation we will be examining in relation to this alignment is **DELPHINUS**, the Dolphin. The name of this constellation originated from Greek times celebrating the lengthy relationship between humans and dolphins, and who were considered the messengers of the sea god Poseidon.

Delphinus has a distinctive shape, its four main stars forming a rectangle known as **JOB'S COFFIN. This is an extremely important and valid point, as it equates to the empty sarcophagus located in the Kings chamber of Pyramid of Khufu, will be explained in detail in volume 3.**

The brightest stars are named **SUALOCIN** and **ROTANEV**, which read backwards, spells NICOLAUS VENATOR, the Latinised form of the name Niccola Cacciatone, assistant and successor to the Italian astronomer Guiseppe Piazzo at Palermo Observatory.

DELPHINUS is composed of the following main stars:
Sualocin - Alpha delphini
Rotanev - Beta (Binary star)

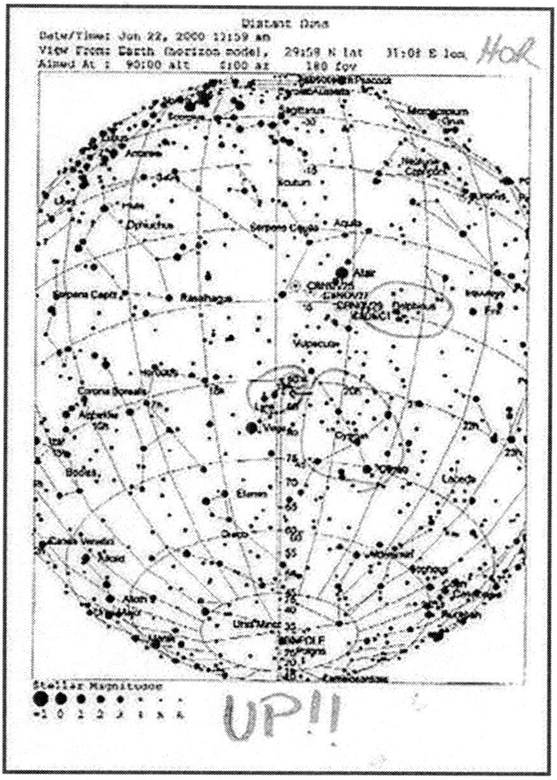

Gamma - Double star
Epsilon
Delta
Zeta

Illustration 5-1
Cygnus constellation

June 21st - 22nd alignment as correlated from the Giza Plateau, in conjunction with the constellation of Delphinus - as above as below.

It may be of note to mention that the date of **June 21st** in the ancient Egyptian calendar is in fact New Year's day and was known as WEPET, occurring in the season of AKHET, the time of inundation of the Nile. This date in the Gregorian calendar is of course the summer solstice, and in ancient Egypt was equated with the rising of Sirius, in turn connected to **OSIRIS**, god of the underworld and resurrection.

It appears that the apex of Khufu aligns with **Zeta** if a line is shot across the apex of Menkaure. Thus, from the apex of Menkaure the three satellites will align as follows.

The east satellite pyramid aligned with the star **Gamma**, the middle pyramid with Delta continuing through to Alpha, while the western pyramid relates to the star **Beta**.

Therefore, as far as we can ascertain, the time of ascension is when the apex of Menkaure, across the tops of the three satellite pyramids thus described, align with these three stars of **DELPHINUS**.

So what is the implication and what, in fact, does this mean?

The configuration (alignment) forms as it is, because this transfer works via an energy circuit. A circuit is created by energy emanating from the star grids, to the earth grid, to the chamber then to the inner earth and back out again. Energy flows in both directions, creating a highly energised balance to all the transfer of bodies. This is very similar to healing methods, when the circuit of the energies from above, through you, and the energy from below also through you, are both sent to the person and then brought back into you. Thus imagine that one is the chamber under the plateau, the synonym of you being the 'channel' applies also.

There is a second alignment through the earth and through the chamber, composing of the stars Lyra and Lesath, basically two cross points with the line that goes through the chamber and the earth, one above the chamber and one below the chamber.

North is M57, the Ring Nebula in Lyra, 2000 light years away, at **+33deg**, and placed between beta and gamma Lyra, yet another sign that this is the correct one.

Thus demonstrating the Earth energy connecting with star energy through the chamber itself.

Upon looking further at the complete set of sky maps, we find several additional correlation and correspondences, which confirm indeed, that this date is correct and give us the exact timing of the start of this phase.

In the early hours of June 22, 2000, at 00:59 a.m., the sky shows us several confirmations and insights.

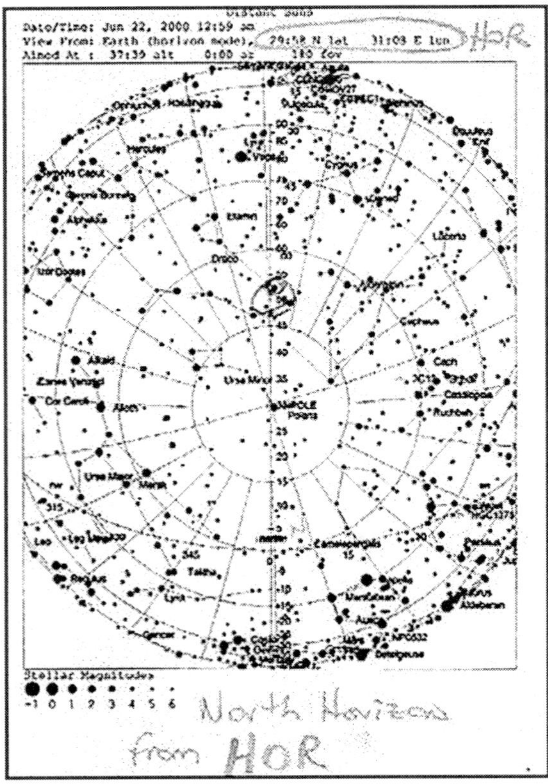

Illustration 5-2
June 22nd 2000 alignment

Looking to the North Horizon from the Giza plateau, we find Draco, specifically the star delta Draco, named Nodus II. **Nodus II** sits at **51.5** degrees declination, which is referenced to the angle of the sides of Khufu.

Also to the North, at **30** degrees declination sits **Polaris** over the North Pole. This **30** degrees declination references to the placing of Giza on the earth, and is always at this position, being the pole star at this point in time.

Now the interesting point of these two stars is that if a line is drawn from them, through the earth, and out the other side, in the corresponding position in the sea near French Polynesia, we discover a corresponding star position. Thus we find at **51.5** degrees declination the constellation Volans, or Flying Fish, specifically **the star gamma Volans**. We can then equate the Flying Fish with Delphinus, that being the ending (flying) of the age of Pisces, where dolphins were the messengers of the gods. Very soon we will not need messengers!

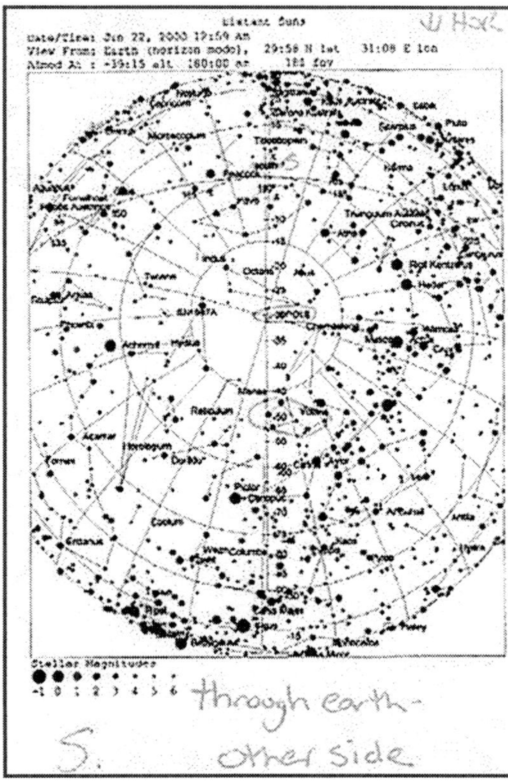

Illustration 5-3 June 22nd alignment – through earth

In addition, at **30** degrees declination straight through the earth from Polaris, is the South Pole, which has no corresponding star. **Although North and South Poles are always there, the fact of North containing a star and South not containing a star, relates to our ascension….at one point we are physically here, and at the corresponding side of the same space and time, we are not.**

If we look straight up into the night sky at Giza, we see **Lyra**. **Lyra** connects straight through the Earth to constellation Canis Major, specifically epsilon Can Maj, named **Adhara**, also known as the Virgins, a double star. Canis Major also contains the star Sirius, well known for its association with Isis. However, at this point we are directed to the Virgins in this constellation, meaning a virgin birth, as discussed in a previous chapter in volume 1 On The Eighth Day.

Turning to the **West**, we find exactly on the horizon the constellation **Virgo**, indicating the Virgin is soon to give birth, as Virgo is setting.

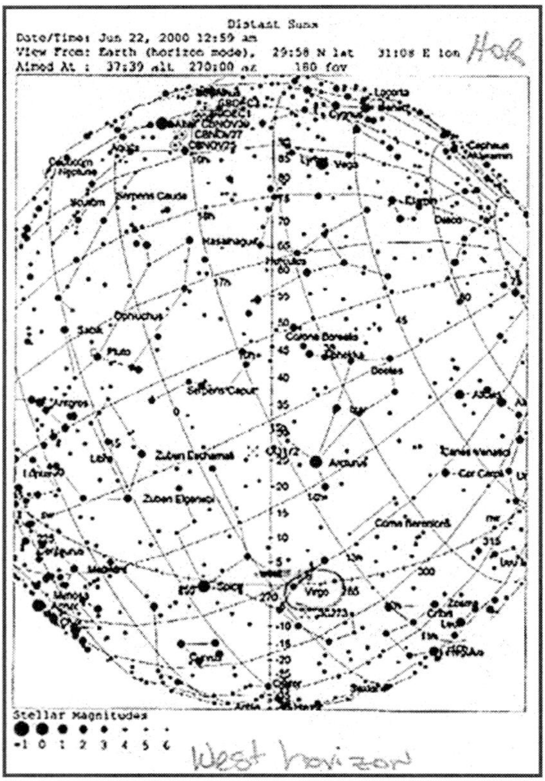

Illustration 5-4 Virgo view from earth

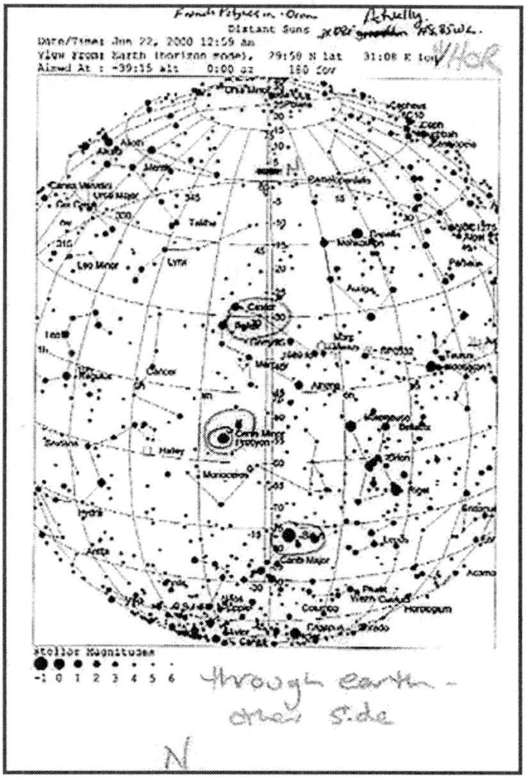

Illustration 5-5 Delphinus alignment

The final alignment we can observe by way of confirmation of this phase is a direct correspondence with the **Cygnus - Delphinus alignment** discussed earlier in this article. Lines drawn directly through the earth from these two constellations connect to the constellations **Orion**, specifically the star **Rigel** (or Giant's Leg), and the constellation Phae Wezn. AND alongside these two constellations runs **Eridanus**, also known as **THE NILE!**

How much more confirmation could we want that this next phase leading towards Ascension starting June 22nd, 2000 at 00:59 a.m.?- relative time .

The next PHASE which (began on JUNE, 22nd) called for the immediate evacuation of all who are prepared, by their right through FREE WILL CHOICE as **"HE"** has promised to do so, **WE** call this promise the **"COVENANT"**. ALL who have lived up to their obligations under the **COVENANT** will have the right to ascend. **THOSE WHO HAVE NOT MUST**

THEN ASK FOR HIS FORGIVENESS. WHICH WE ARE SURE HE WILL GRANT. For **"HE"** would never knowingly or unjustly, forsake any of his children.

There will be a warning prior to the event for all people who are leaving, however, those selected to go will only know minutes beforehand that this is to occur. It is believed that when they "feel" that call, (the call being a disoriented feeling) they should immediately relax as fully as possible. Others involved will carry out the remaining work necessary.

There is a welcoming committee in place to help the people re-orient themselves once aboard the large spacecraft. Those selected will be unable to bring anything with them, apart from what is on their person at the time. We have been advised that they will be made very comfortable for the duration though out the journey and will eventually see how irrelevant the physical is anyway.

June 22nd was the start of a **"window of opportunity"**. It is anticipated that the take-up will not happen all in one day. Those that are in isolated areas will be taken without too many problems. Others will have to be taken when other events are happening simultaneously, whether that be war, weather or earthquakes, etc. As the EARTH is now about to undergo a series of transformations, these changes will be occurring in preparation for the POLE SHIFT. During these EARTH adjustment event's, millions of people will SEEMINGLY be presumed dead. THIS however will not be the case, as this will be the time when those going will be removed and taken aboard the waiting ship's.

In other words, the process will be carried out when these people will not be missed immediately. **The ones left will have ample time to make their choice at a later date..**

The method of 'transportation' set in place would basically be described as a **'Jacob's ladder'**, which will remain in place until the battle commences. Once the battle lines are completely drawn and everyone is in place (**after the birth in December**) the ladder will be closed.

This process of 'ascension' will continue until all those who wish to go are taken, or until the window closes due to other foreseen events. The next coming months will witness much travail on the earth and many changes will be occurring to the earth. **.**

Although there is some evidence to support the LAST EXODUS, it is as often left to the individual to make his or her own decision, to place oneself into "HIS" care. For this requires FAITH in "HIM".

One may wish to give thought to a number of cities that have been discovered, such as ANGKOR WAT, where the city remains were found intact (only the people missing or disappeared without trace). There are many more, whether Aleatian, Mayan or others, all found totally intact.

Chapter 6

THE NIGHT OF THE RED SKY

In our previous chapter entitled "The Line Has Been Drawn", we explained the stellar alignments as they occurred over the Giza Plateau, Cairo on June 22nd 2000. In addition, we offered an explanation for this crucial event (**Ascension**) and the consequences involved for all on this planet.

In this chapter, we now offer further evidence, the necessary essential details and an overview of the time line of events that led up to December 21st 2000 relative to our present time line.

What was so important concerning the Dec 21st date?

In a nutshell, certain solar events (some having been recently manipulated in order to alter the sun's 'natural' evolution) will lead to the birth of a new planet from our sun. Thus equating the 'myth' of **ISIS (Mary)** giving birth to a new planet.

However, not only will we (for those remaining on planet Terra) witness the birth of a new planet into our solar system, but also over the next few months Terra will experience many 'earth changes'. This may be in the form of severe weather, earthquakes and volcanoes etc. as she reacts to and accommodates for the evolutionary process that is taking place. These events will eventually culminate in a POLE SHIFT.

We will now endeavour to explain these events, the background scenario currently taking place and the consequences for planet Terra.

A serious contender in the instigation of anomalous activities surrounding this scenario is the Brookhaven Laboratories, the home to the infamous

Relativistic ion collider, (http://www.bnl.gov) where attempts are being made to alter the earth shifts by directing energy paths thus preventing earthquakes from occurring on the continental US shelf. Further details of this scenario we will discuss later in the chapter.

THE SUN

Our star, (the sun) is the life-blood for survival of planet earth·and source of heat, light and energy. According to the traditional and government sources, we are informed our sun is an "ordinary star similar to millions of others in the universe." Sources tell us it is an, 'energy machine' producing enough energy in one second to provide the USA, for example, with enough energy for the next 9,000,000 years!

According to the same source, (NASA) as above, in about 10-20 billion years the surface of the sun will expand enveloping the inner planets of our solar system, including earth, and thus creating a red giant star. Eventually, because of its size, it is expected to contract into a smaller cool star known as a white dwarf.

However, we have evidence to suggest the contrary, including the fact that our star is now a **Helium star** (since 1991) and in its present condition highly unstable, ultimately being given approximately only 1,045 years of extended life. That is, in 3045, our sun will be going **SUPER NOVA!**

It seems that in 1962 the sun lost its polarity and is assumed to have reached half-life, based on telescopic observations made at the time from Aerospace stellar facility Simi Valley California. Some years later **SKY LAB** was launched to observe the solar anomalies occurring at the time. In 1978 SKY LAB eventually plummets to earth in the region of western Australia. The next probe to be launched was **SOHO** (The Solar and Heliospheric Observatory) and **Ulysses,** verifying the fact that no polarity existed on the sun and it was in fact all helium.

In November 1997, for those following the SOHO satellite images observed large anomalies orbiting the sun. These 'anomalies' can still often be observed today and have given terns such as 'Sun cruiser' and in one of our chapters in Volume 1, **THE FINAL COUNTDOWN**, we named the anomalies 'Plasma carriers'.

It transpires that these anomalies are in fact gigantic Extra Galactic spacecraft, which came in towards the sun and possibly delivered two Venus sized missiles into the sun, causing huge flare explosions.

The comment by NASA at the time is summed in the headline…*Two comets slam* into *the sun.* They also added that SOHO was dislodged and located three million miles from its original location. Indeed, in May 2000 we were witness to further 'missile' activity towards the sun.

We quote a brief by Kent Steadman of cyberspaceorbit.com regarding a solar event and the sun regaining its polarity in 1998:

> The WWW also went wild and stated: (SKY WATCH) "What are they trying to do...Blow up the SUN"?...I had just Joined the "NET" at the time and took it upon myself to give a "Scientific" explanation of WHAT had happened and WHY as well. The WHY was the above history; the WHAT dated back to 1951 at Berkeley...Dr Pauling, the Nobel Laureate at Berkeley (My Chemistry GURU) had been doing "end-of-Elemental relationships" with the newly installed Cyclotron. TWO NEW ELEMENTS were discovered: They were called "COSMIUM" (as it, being the "end-of line" element. discharged "Cosmic Particles", (that also disintegrated everything in their path) and the next EOL was one named ISOLIUM. (able to STORE ELECTRONS as does a "capacitor", only beyond imagination) Each had a 300 year "half-life". We made a Self-powered" PNP Transistor from Cm, Is, and Pb. It was capable of 1KW of pwr for the entire half life at 150 VDC output. (in a Lipstick tube yet) The battery we made was the size of a PEA for the same spec's! The NEW Elements have since been renamed: Berkelium, and Californium. The project being BEYOND TOP SECRET (and still is) proved to be the deterrent that stopped Kruschev in 1962 in CUBA as we had made a PLANET-BUSTING BOMB from the newly discovered elements. "NATURALLY?" Sheeesh. NOW that took imagination...wot?

If one should contemplate the idea of moving our planets around the solar system, without any forethought to the understanding of exactly how moving one element, such as the earth, could alter the electro- magnetic flow, then the below article which was released on June 10th 2001, may be just ONE of the reasons why this up-coming battle would be fought.

It is this irrational thought process and lack of understanding of how our planetary grids are all interwoven, that seems typical of an elite few. As if one moves and therefore alters this finely interwoven grid structure, then there would be a strong possibility of the vaporization of a star system on the other side of our Galaxy. It is because of this that no intelligent life form out there in the cosmos would wish to be our neighbors, and who would blame them?

NASA aims to move Earth

Scientists' answer to global warming: nudge the planet farther from Sun

Robin McKie, science editor
Sunday June 10, 2001
The Observer

Scientists have found an unusual way to prevent our planet overheating: move it to a cooler spot.

All you have to do is hurtle a few comets at Earth, and its orbit will be altered. Our world will then be sent spinning into a safer, colder part of the solar system.

This startling idea of improving our interplanetary neighborhood is the brainchild of a group of Nasa engineers and American astronomers who say their plan could add another six billion years to the useful lifetime of our planet - effectively doubling its working life.

'The technology is not at all far-fetched,' said Dr. Greg Laughlin, of the Nasa Ames Research Center in California. 'It involves the same techniques that people now suggest could be used to deflect asteroids or comets heading towards Earth. We don't need raw power to move Earth, we just require delicacy of planning and manoeuvring.'

The plan put forward by Dr Laughlin, and his colleagues Don Korycansky and Fred Adams, involves carefully directing a comet or asteroid so that it sweeps close past our planet and transfers some of its gravitational energy to Earth.

'Earth's orbital speed would increase as a result and we would move to a higher orbit away from the Sun,' Laughlin said.

Engineers would then direct their comet so that it passed close to Jupiter or Saturn, where the reverse process would occur. It would pick up energy from one of these giant planets. Later its orbit would bring it back to Earth, and the process would be repeated.

In the short term, the plan provides an ideal solution to global warming, although the team was actually concerned with a more drastic danger. The sun is destined to heat up in about a billion years and so 'seriously compromise' our biosphere - by frying us.

Hence the group's decision to try to save Earth. 'All you have to do is strap a chemical rocket to an asteroid or comet and fire it at just the right time,' added Laughlin. 'It is basic rocket science.'

The plan has one or two worrying aspects, however. For a start, space engineers would have to be very careful about how they directed their asteroid or comet towards Earth. The slightest miscalculation in orbit could fire it straight at Earth - with devastating consequences.

It is a point acknowledged by the group. 'The collision of a 100-kilometre diameter object with the Earth at cosmic velocity would sterilise the biosphere most effectively, at least to the level of bacteria,' they state in a paper in Astrophysics and Space Science. 'The danger cannot be overemphasised.'

There is also the vexed question of the Moon. As the current issue of Scientific American points out, if Earth was pushed out of its current position it is 'most likely the Moon would be stripped away from Earth,' it states, radically upsetting out planet's climate.

These criticisms are accepted by the scientists. 'Our investigation has shown just how delicately Earth is poised within the solar system,'

Laughlin admitted. 'Nevertheless, our work has practical implications. Our calculations show that to get Earth to a safer, distant orbit, it would have to pass through unstable zones and would need careful nurturing and nudging. Any alien astronomers observing our solar system would know that something odd had occurred, and would realise an intelligent lifeform was responsible.

'And the same goes for us. When we look at other solar systems, and detect planets around other suns - which we are now beginning to do - we may see that planet-moving has occurred. It will give us our first evidence of the handiwork of extraterrestrial beings.'

In addition to this available evidence, if one were to read Dr Noel Huntley's work [1] you will see that he offers the explanation that the Galactic Federation altered the basic polarity of the sun to allow it to maintain the integrity of the sun's planetary system.

In recent times we have been experiencing increasing solar and geomagnetic activity emanating from the sun as the charts and satellite images show at Orbit, e.g. solar X-Class flares. In addition, much speculation and debate surrounds the scenario of Comet 76P West-Kohotek-Ikemura and its close encounter with Mars

The recent missiles fired into the sun. will have started an internal electrical reaction within the sun itself. This reaction will begin to be absorbed at the core level at which time the core will begin its next cycle at its stage of growth. This will continue until the great alignment of **December 21st 2012**.

While we were researching this particular section of the book, new evidence appeared in research utilising SOHO spacecraft [2] that the surface of our sun is covered with long-lasting depressions and humps that are very similar to features on the Earth's oceans.

In conclusion the authors note… "That the Sun is covered with "hills", each about 100 metres (330 ft) high and 90,000 km (56,250 miles) apart. If the Sun was scaled down to the size of the Earth, these features would appear as small bumps only a few centimetres in height." It is thus possible to predetermine the exit points through which the **NEW PLANET** will exit, this is calculated by knowing (and understanding) the **KEY PRESSURE POINTS** of this complex **GRID**.

ALIGNMENT DETAILS

The June 22nd and December 21st alignment 2000

June 22nd - portal locations

To view the full alignment details, including the star charts, please refer to the relevant section in our previous chapter, The Line Has Been Drawn.

The portal alignment on June 22 is directly over Giza, it is at this location where the ships will enter, following the grid lines to their respective destinations.

Once the ships are through this portal into our atmosphere the portal will close. This portal will remain closed, (baring unforeseen events) until all whom have chosen to leave have had the opportunity to do so.

The ships must, however, must have exited prior to the "BIRTH EVENT" on December 21st 2001. ALL those who remain here will do so until after the Battle (written as Armageddon).

This procedure is all carried out via the Purple Grid, (blue over red) This portal allows the ships to enter the Purple Grid, where navigation is carried out within this Grid from that location.

THIS window of "OPPORTUNITY" will last for a maximum of (slightly under), six months. During this six-month period MANY EARTH CHANGES will occur. IF you had made your choice and your name has been recorded, and find yourself in a situation during one of these Earth Changes Events, just try to relax and they will remove you from the situation. The locations/areas will not be totally unknown at this time, as a number of people have been given clear VISIONS of these events, including the areas in which they will occur. WE realise that there is a certain amount of FEAR attached to such events, BUT again we must stress that you do not fall victim to the fear.

We suggest that when one gets that feeling just relax.

THE BIRTH EVENT

In our chapter *On the Eighth Day*, Volume 1, we mentioned that Isis/Mary revealed herself standing in a strong wind and it appeared she was Pregnant, we believed that she was trying to show us that she was pregnant with Ephraim. We do not have verbal communication with Isis/Mary, however, shortly afterward she re-presented herself, but this time in her company was a small boy. The boy presented himself, as having short dark hair, smooth dark skin and wearing a white tunic, while around his lean waist was a Gold and Burgundy braided band. In his Blue-Green eyes you could make out that his energy is similar to Anubis, but less humorous. When asked his name he replied Ephraim. Upon his head was a small crown, and in his hands, attached to a red string, he held a medallion, similar to a Gold coin. He appeared rather shy and did not want to be seen, disappearing into a shadow as he left.

It was at this time that we knew that Isis/Mary had presented herself being pregnant to make us aware of the forthcoming birth event, the birth event that we have since confirmed is that of our sun giving birth to a new Planet. In addition, the travail, which Thoth/Joseph informed us of, was in relation to the changes (electro-magnetic) our solar system would be going through in order to accommodate this New Planet.

See following images.

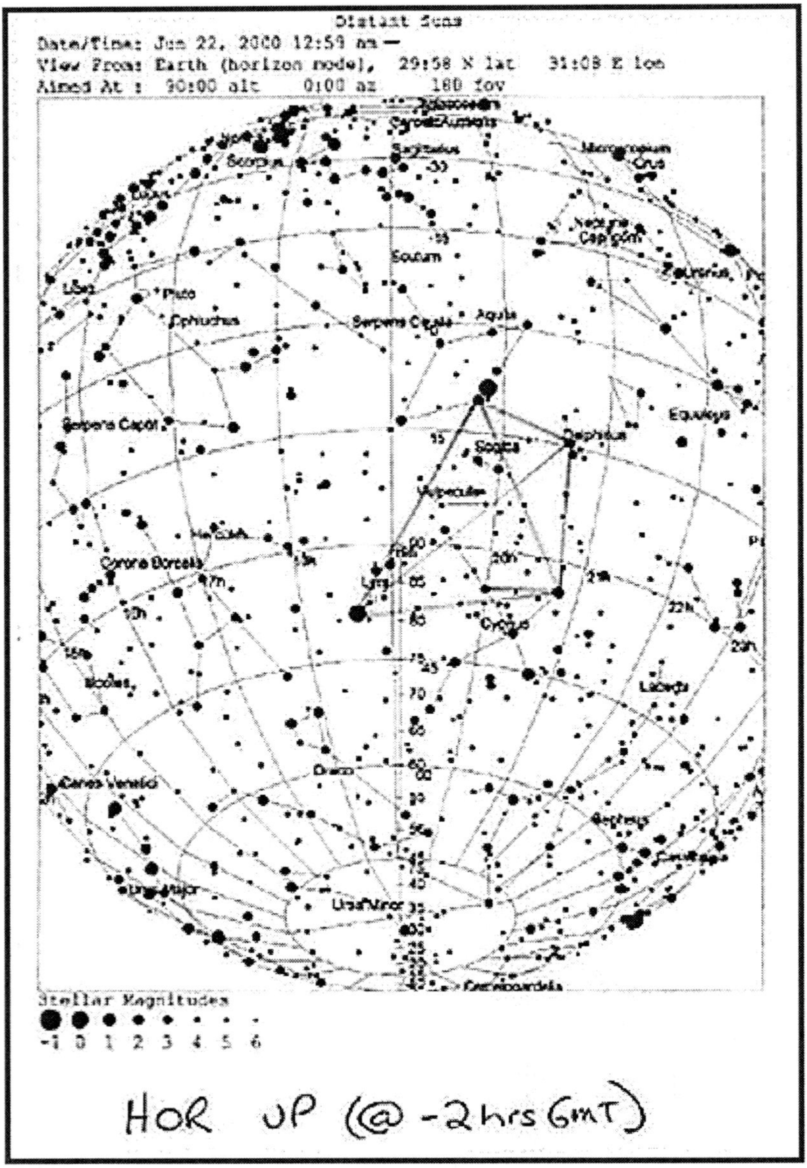

Illustration 6-1 June 2000 alignment

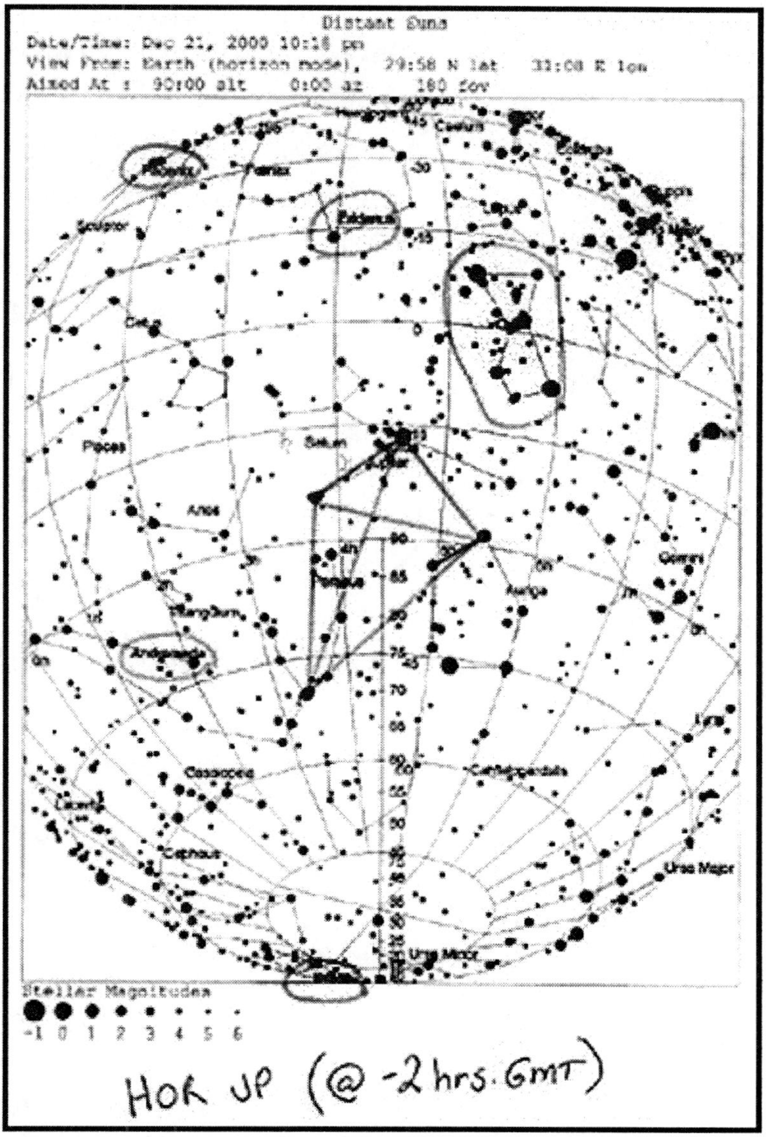

Illustration 6-2 December 21st 2000 alignment

The major players arriving for the December 21 event at Giza.

Directly to the South on the horizon will be Reticulum, home of the **Zetas** or Greys, while to the North on the horizon will be Draco, home of the **Draconians**.

These two groups are here to watch the events as they occur, as eventually their ships will be entering for the battle (written as Armageddon) and joining forces with Esau.

The Galactic fleet, led by Michael, departed the Pleiades on May 28th 2000 and will be here to intercept them. When the poles change, our present grid will no longer be functional consequently the Draconians and their followers will be able to enter this space.

At this point we will still be held within a Time Grid, however the artificial protection grid, will be thrown out of synch.

Directly to the West on the horizon will be Aquarius. Aquarius is the water of life and is setting, so that Leo and our Sun can do their work. When the Age of Aquarius rises in the future, it will be to extinguish the Fire on Earth and subsequently bring Life.

Directly to the East at 15 degrees North horizon will be Leo, rising. Leo is the face of Fire, relating to the birth - The rising. While straight through the earth from the vortex, will be Scorpio, the Eagle, representing transformation!! Eagle flies highest of all birds and so also represents the rising of the people of the Covenant.

BROOKHAVEN

At Brookhaven National Laboratories in New York, attempts are being made at replicating the "Big Bang", the traditionalist view of how our Universe was created. At the **RHIC** (**Relativistic Heavy Ion Collider**), heavy ions are "shot" through two tubes, the Blue and Yellow tunnels, to collide at high frequency in effort to recreate this effect. In so doing, the **Electro-Magnetism (E/M)** surrounding the ions is affected.

What has been observed since the RHIC has been conducting these experiments is that the E/M of the surrounding environment is also affected. This can be seen in the weather radar maps, whereby the air over the facility is "heated up", creating an E/M field over the surrounding area. Seen in this heightened E/M field over the area when BNL has the collider up to speed, is a beam connection in the direction of HAARP. The combination of the E/M from BNL and the output waves from HAARP, result in the RHIC energy being sent along grid lines to their targets, as shown in Julie King's article HAARPSTRINGS in Volume 1.

It is believed HAARP on it's own can create E/M fields over various targets, resulting in earthquakes, weather changes and physiological symptoms. This can be carried out directly through the grids or sent to satellites and back down to the target. It is also believed that HAARP could be used to create holograms in the sky as part of PROJECT BLUE BEAM)

The effects of HAARP have been seen for some years now via the weather maps, as shown on "Flash Radar". The addition of RHIC to the scenario offers new aspects as to what can be affected.

By combining the E/M from RHIC and HAARP emissions, it is quite possible to create a hole in our atmosphere, in fact, a hole in the grid itself! If this were to occur, any ships would be able to enter our atmosphere, as described in our previous work, On the Eighth Day. This might also cause similar effects to the Philadelphia Experiment, whereby a hole is created in the Time - Space Continuum.

In addition, it is apparent that they do not want the public to know about the imminent Pole Shift! For indeed the combination of these two technologies have been used to alter the tectonic plate shifts of the Atlantic, thereby preventing plate shifts which could create a shifting of the Earth herself!

As we mentioned previously, on May 2nd 2000 an interesting piece of news was revealed, that of scientists discovering cracks in the ocean floor off the east coast of the USA. It as revealed that these cracks along a 25-mile stretch of the continental shelf could trigger a tsunami sending 18-foot waves towards the Mid Atlantic States. The news as reported by the Associated Press on May 2nd, Falmouth, Mass, where the New Millennium news site issued an abbreviated report. However, the full article may be read in the May 2000 edition of the journal ***Geology*,** reference article No.407

This event seems to attest to the instability of the continental shelf, illustrating yet a further symptom of planet earth's reaction to its growth stage towards the **2012** date.

For the time being however, there will be no such scenario, as the Purple Grid holds Terra in her correct Time - Space co-ordinates. Terra will be held at these co-ordinates until the continuum flow reaches the correct Time - Space co-ordinates for the battle of Armageddon.

THE ALIGNMENT CONFIGURATION FROM THE GIZA PLATEAU

The basic structure in our alignment is that the **Southern cross**, or **Crux**, is located directly over the Giza Plateau, and it is this location that the vortex is situated.

By rotating our Plateau model 120 degrees in a south to north direction (clockwise) and applying the model to the star chart, the result is a mirror image - as above so below.

Taking the configuration from the base of the cross, connected to the apex of Khufu, the following correlation is found;

Khefre is seen to be represented by **Aldebaran in Taurus**, Menkaure by the Pleiades and Khufu by **El Noth in Taurus.** Isis, the satellite pyramid of Khufu, is represented by **Hasselah Auriga**, the Sphinx by **Epsilon in Perseus** and the Valley temple by **Nu in Perseus**. While the location of HOR, as determined in our previous articles, is represented by **Algenib in Perseus.**

NIGHT OF THE RED SKY:

Many will know of the famous **Fatima** (Portugal) prophecy where it is foretold of **'Red Skies'** will be witnessed prior to an event of great proportions, but how many know of a similar and equally profound vision by an Apache in the 1920's.

We now refer to the article in Nexus magazine [3] where a series of prophecies, or warnings, were made by an Apache wise man in the 1920's, foretelling the death and destruction for mankind.

> **It was a vision of destruction of man. But man was given four Warnings to that destruction, two of which gave man a chance to change his ways and two of which would give the children of the earth time to escape the creator's wrath.**

The **first sign** was one of a famine and mass disease born of monkeys, drugs and sex, this was taken to mean Aids, in addition to wars in the cities as a result of this scenario.

The **second sign** was that of a barren earth landscape, animals dying where the sun appeared much larger and more intense...

> **A floating garbage of oil and dead fish........floating bodies Of dolphins accompanied by tremendous upheavals of the Earth and violent storms.**

Bear in mind the part that the constellation of **DELPHINUS** plays in the **June 22nd ALIGNMENT!!!**

In recent times we have read of many reports of Dolphins and whales being beached in large numbers Holes in the sky would be the second sign of destruction of man, and it was said that from this point forward man could no longer heal the earth with physical action.

> **These holes are a direct result of man's life, his travel and Of the sin's of his grandfathers and randmothers......... And the grandchildren will be faced with a choice - to continue following the path of destruction, or to move back to the philosophy of the earth and a simpler xistence.**

The **third sign** is when the sky turns red, even the stars shine red...

This, then, is the third sign, the night of the bleeding stars It will become known throughout the world, for the sky in all lands will be red with the blood of the sky, day and night. From this point there will be no hope - life on earth as man Has lived it will come to an end, and there can be no turning Back, physically or spiritually.

At this point, it may be useful to briefly look at the incredibly accurate earth changes predictions of one Michael Scallion, who refers to the so-called **Blue Star**. When this star is close enough, angular forces will be set up between Sirius B, Arcturus and our sun. This alignment along with this new Blue Star will trigger a polar shift in the sun. Scallion predicts that our sun will re-position itself internally, thus altering its field, immediately transferring that change to the earth's core, and all other planets in the solar system. Consequently, the earth will change its magnetic field and the north magnetic pole will shift its position. .

This so called "Blue Star" is known to scientists as the Hale-Bopp comet, and Scallions' name for the comet comes from the old Hopi legend about the dancing Blue Star Kachina. Several other prophecies have mentioned a new sun at the end of the age including Mary's messages, the Ramala readings, and those of Nostradamus.

To return to the visions of Stalking Wolf, the **fourth sign** is of a great famine and a disease that kills the masses

Man will be destroyed, his cities in ruin.

The **CHILD** delivering this message was **"EPHRIAM".**

This will last for a period of ten years, bringing us to the year 2011 (Dec. 21st 2011) This period will denote the completion of the Battle (Armageddon), and, like Stalking Wolf, there will be one year to

look back and truly absorb what took place, and to learn and understand that the SPIRIT of MAN and the SPIRIT of EARTH must learn to live in HARMONY...On December 21st 2012 TIME will then be RESET...

(Endnotes)

[1] http://www.users.globalnet.co.uk/~noelh/index.htm

[2] (http://spaceflightnow.com/news/n0006/13sunbumps/

(1) [3] Nexus http://www.altnews.com.au/nexus/RedSkies.html , Volume 7, Number 1 (December 1999 - January 2000).*Night of the Red Sky The Prophetic Vision of 'Grandfather'*

Chapter 7

UNIVERSAL CONSCIOUSNESS

It is important to comprehend here, that the act of 'circumcision' (The sign of the Covenant) is not really the medical term, as we know it today. This 'sign' applies to all mankind, not just the males of this planet. The shedding of the foreskin is about the personal control of the suppressed **SPIRIT** in or within us, a time to renew our faith in the harmonic workings of the spirit as a whole. That is the **Universal Consciousness**.

That time is known in the Mayan calendar as the 13th Baktun, where one must shed or sever the reliance on only the physical world, or the underworld.

Having taken on board this explanation, we move forward.

We mentioned that as soon as possible after the June 22nd alignment, the 'ascension' or exit, commenced and will continue as long as that window of opportunity exists. We also offered an explanation as to the physical symptoms experienced prior to leaving and what signs to look out for.

Many have enquired who will go, and that will be dealt with in volume 3, however, it must be borne in mind that the 'dye was cast' aeons ago, the book was written and all who are destined to go will have been recorded in that 'book'. However, there is still that option of Free will choice, one only has to ask! This is not the first time we have been asked to make this "free will choice", and it may not be the last time, however, we truly believe that one-day it will be so. On each and every occasion we all pray that it will be the last time.

Some have asked why Alien races from distant worlds would come here to fight a battle, in our defence. This is not so very hard to understand as The Galactic Federation is made up of a number of diverse races of people. All of who had to face this situation at one time or another in the struggle to be free, and remain free of those who would enslave them, and take away their right to have the right to exercise their own free will choices. In the beginning the battles were being lost because they were very strong, on sheer numbers alone, where most worlds often found they had no assistance in their struggle. It was then that the Galactic Federation came into being. The war rages on as these pockets of Slave Merchants are located and systematically confronted.

It is a big galaxy out there, with many dimensions and many time frames, and these time frames, when major confrontations may occur, can actually be calculated aeons before the event itself. Thus it is during these times that the dark side will infiltrate into the positive side of space. That is, Dark versus Light.

When the call is given for assistance the federation responds quickly. As we mentioned in our previous chapters who can ask for this assistance? (see chapter on **The Crystal** in Volume 1). The one whom the blessing has been entrusted to, in our case (Terra), that is in the hands of Ephraim.

When we were first conceived, it was of the flesh then the spirit entered the flesh. In other words the conscious versus the subconscious minds, where the Spirit is connected to the Universal Consciousness and the conscious mind to the physical world. The two, however, are capable of working in harmony. The sad part however, is when the conscious Mind fails to listen to the SPIRIT within, thus creating a severe imbalance. When the conscious mind places itself above the Universal Consciousness it creates an imbalance. IF THERE IS NO BALANCE THERE CAN BE NO HARMONY.

So what does the 'mark' of circumcision really mean in this context? We must break loose of this bondage that holds back the spirit from the harmony of the UNIVERSAL CONSCIOUSNESS. We are of a positive12 strand DNA, whereas our suppressors are of the Reptilian negative strand DNA. THEY exist here within our time frame, but in a different Dimension. They literally tap into your Life force and live off our very being, they are also very weak minded and have lost their individuality. They have lost their contact with the Universal Consciousness. However, do not misunderstand, not all reptilian life forms are like this, it just happens that the ones who hold us hostage are. In volume three we discuss this in greater detail in the chapter dealing with Esau.

The circumcision process is when you TAKE BACK CONTROL OF THE SPIRIT WITHIN YOU, AND THROUGH EXERCISING YOUR RIGHT TO FREE WILL AND PLACING YOUR SPIRIT BACK INTO HARMONY WITH THE UNIVERSAL CONSCIOUSNESS.

It is your responsibility to instruct them to leave go, to leave you alone and not to bother you again (if that is your free will choice). Occasionally, one must repeat this process, for they are like leeches, one may leave and another may attach itself.

We explained in Volume 1, via Genesis, the lineage of our planet Terra, where earth represents Adam, our sun Abram and its lineage traced back through the Pleiades from the star Alcyon. Eve is extracted from the sun's ribs and thus the earth is born from the sun. We then further explained the two remaining lineages' of Haram (Chaldean) and Nahor (Seth, or reptilian lineage).

15:13

> Then the LORD said to Abram, "Know of a surety that your descendants will be sojourners in a land that is not theirs, and will be slaves there, and they will be oppressed for four hundred years.

This is the fourth generation, and is completed once earth reaches the present location of mars at 1000 or 'zero' degrees. As in 'revelation' the '1000 year reign', the 13th. Baktun, 13 degrees to the inside from the point of origin.

THE EIGHTH DAY

Heaven and earth were made in 6 days, the 7th rested, thus the 8th. was the first day of existence, this is where the reptilian form is 'lost' (the skin is shed)

17:9

And God said to Abraham, "As for you, you shall keep my covenant, you and your descendants after you throughout their generations.

17:10

This is my covenant, which you shall keep, between me and you and your descendants after you: Every male among you shall be circumcised.

17:11

You shall be circumcised in the flesh of your foreskins, and it shall be a sign of the covenant between me and you.

17:12

He that is eight days old among you shall be circumcised; every male throughout your generations, whether born in your house, or bought with your money from any foreigner who is not of your offspring,

17:13

both he that is born in your house and he that is bought with your money, shall be circumcised. So shall my covenant be in your flesh an everlasting covenant.

17:14

Any uncircumcised male who is not circumcised in the flesh of his foreskin shall be cut off from his people; he has broken my covenant."

In **Genesis17:18** we read:

And Abraham said to God, "O that Ish'mael might live in thy sight!"

This is the point where this race shed their skin, Abraham at '99' turning 100 Ishmael at 13 (degrees), which equates to Baktun, the end of the cycle.

17:23

Then Abraham took Ish'mael his son and all the slaves born in his house or bought with his money, every male among the men of Abraham's house, and he circumcised the flesh of their foreskins that very day, as God had said to him.

17:24

Abraham was ninety-nine years old when he was circumcised in the flesh of his foreskin.

17:25

And Ish'mael his son was thirteen years old when he was circumcised in the flesh of his foreskin.

17:26

That very day Abraham and his son Ish'mael were circumcised; 17:27
and all the men of his house, those born in the house and those bought
with money from a foreigner, were circumcised with him.

The circumcision as relates to Genesis, symbolizes the covenant made between Yahweh and his people. This symbol is often found in much 'older; civilizations, but in Biblical Hebrew it literally means to cut a covenant, the cutting in two of a sacrificial animal. While Deuteronomy talks of "Circumcise therefore the foreskin of your heart" **10:16**, "and be no more stiffnecked". While Paul comments "neither circumcision nor uncircumcision is of any avail, only faith that works by love". **Gal 5:6**.

THE BROOKHAVEN SCENARIO

The United States' Department of Energy's Brookhaven National Laboratory New York home to the Relativistic heavy ion collider (**RHIC**). Do the scientists and the team at Brookhaven know what they are really playing with? By the results showed thus far we suggest no.

Do they know who is pulling the strings and what the actual agenda is? Again we suggest no.

In the summer of 1998, this news item was released

MAGNETS COMPLETED FOR BROOKHAVEN LAB'S NEW COLLIDER

About 300 invitees assembled in Brookhaven National Laboratory's Relativistic Heavy Ion Collider (RHIC) Magnet Facility to celebrate the successful completion of a seven-year, $165-million magnet project, requiring 70 million feet of superconducting wire and 929,500 technician-hours.

Funded by the U.S. Department of Energy (**DOE**), **RHIC** - Brookhaven's new collider - will use 1,740 superconducting magnets to guide beams of subatomic particles around its 2.5-mile underground tunnel when it starts operating in the summer of 1999.

Taken from http://www.pubaf.bnl.gov/pr/bnlpr100698.html

The general idea of the ion collider is that of detecting head-on collisions between gold nucleiin the Relativistic Heavy Ion Collider (**RHIC**), the world's newest and biggest particle accelerator for studies in nuclear physics.

> While the beams have been in collision mode since the weekend of June 10, the first spectacular images of particles streaming from a collision point — the definitive evidence the scientists were waiting for — were produced by the **STAR** detector last night at 9 p.m. High-energy collisions were also seen by the **PHOBOS** detector early this morning.

http://www.rumormillnews.com/cgi-bin/config.pl?read=3660

Against the backdrop of many scientists who argue against the feasibility of such experimentation, such as interfering with the cosmic organisation of the universe, and producing unknown effects of attempting 'Black hole' replication, this statement was made...

> The director of the laboratory commissioning the machine says there is "no chance" of the atom-smashing experiment causing a disaster, such as a black hole that would devour the entire Earth.

From news section at http://www.sightings.com/politics4/rep.htm

June 10th saw the 'firing up' of the experimentation and following the activation of Brookhaven, much radar activity was observed around the Long Island area.

The first images of these atom collisions were released on the 14th June and with it the accompanying news brief:

Big Bang machine gets down to work

June 14 — The first spectacular images of atoms smashing together at near-light speed were released Wednesday as part of an experiment scientists say will eventually generate a state of matter that existed a millionth of a second after the Big Bang. A milestone in physics, the researchers' effort has also generated heated debate and several doomsday theories — including one that argues the experiment might release particles called "strangelets," which could gobble the globe.

However, serious implications and repercussions will no doubt begin to materialize as to the true purpose of this machine.

Elfrad, picked up some serious anomalies related to Brookhaven

Indeed, the comment made by Julie King at the time seems to sum up the beginning of something unusual in the making:

If this [HAARP data] was RHIC then we are in big trouble...it knocked out our electricity here big time...was a very strong hit. The battery backup clocks, etc., went out too, the batteries didn't click in, like maybe they were discharged for that time period? Meanwhile, I've had a splitting headache ever since.

Shortly after this event, anomalous debris field was observed on the SOHO solar satellite.

But what does it all actually mean?

It appears, as intimated previously, that not all is as it seems. That is, what exactly ARE they doing and where does it fit in with our lives on this planet?

Bear in mind also, that many of the scientists working on this project are working there on a "need to know" basis, where the full agenda is hidden, and only available to an elite few. So who is pulling the strings and for what purpose?

We believe that this project is an attempt to perfect the ultimate weapon of destruction, where the sole purpose is to only kill and annihilate all living organisms, while at the same time leaving the physical assets remaining!

We described in our previous chapters On the Eighth Day and Exit us the purpose of the protective electro magnetic global grid encircling our ionosphere. Attempts are being made to destroy, or at least, punch a hole in this grid in order for certain entities to enter into our atmosphere, for an agenda that we also explained in On the Eighth day. However, by working for one (Esau) who dies not care as to what state the planet is in as long as he 'reoccupies' the same, scientists do not realize the highly serious nature and implications of their actions.

Is this a case of the Atomic Bomb scenario of World War two all over again, only on a global scale? Too many related events have been observed over recent months to rule this possibility out. E.g. Los Alamos fire for one.

The downside of this scenario is that if allowed to continue and if they hit the wrong/right level of emission (density and frequency levels), all organisms contained between the confines of the field would be annihilated. Thus the idea that even the smallest organism can be mountaged through time is nought...Each and every organism is made up of smaller parts, and by the time they hit the frequency, all would be gone before they could gain any control, even if they could.

On the upside, the grid now in place (the purple grid) around our ionosphere, absorbs and equalizes all energy, no matter what that energy is. In addition, many of the followers of Esau have remained on this planet, so it is doubtful (but not totally out of the question) that he is going to eradicate his own also. It is also a case of being slightly too optimistic in thinking that may be the battle may be in his favor!

As at June 22nd, alterations to the vibration of this grid will have been activated, this in itself will confuse the operating systems, such as at Brookhaven, at least for a time.

The important point here is to bear this probability in mind, especially to those who are operating the systems at Brookhaven and are totally unaware of the **REAL** agenda.

THE CHEMTRAIL CONNECTION

In our previous section Exit US, we commented on the serious and anomalous activity of Chemtrails, as a result of this global aerial spraying many people have been experiencing a number of diverse flu-like symptoms, some more serious than others. A few researchers have taken it upon themselves to analyse the 'fall out' composition of this aerial residue, which in many cases has deposited on the ground as cob-web type filaments.

To cut to the chase here, those particles are highly charged (electrically) and have been deposited and saturated through practically every living organism on this planet!

What has this to do with Brookhaven? **PLENTY…**

In one word –Montauk, the infamous location of those bizarre experiments, coupled with the Philadelphia experiment, concerning time travel. Originating out of this project we now illustrate with a diagram of a schematic of part of the experiment.

http://www.ambilac.com/images/part_xii_images/schema_2.jpg

This shows that the two power sources are comparable to the blue and yellow tubes at Brookhaven. The release point being Brookhaven and the Booster HARRP, (HIGH FREQUENCY ACTIVE AURORAL RESEARCH PROGRAM)(1)

HAARP fires directly up the grid, while Brookhaven on the other side, fires it out towards the grid. This then is contained between the grid and the charged particles which they have been spraying and blanket covering - THANKS GUYS. This now offers a build up within the grid and at a wrong density and frequency could result in total annihilation of all organisms existing on that particular bandwidth!

A FINAL NOTE TO ALL THOSE AT BROOKHAVEN

A WORD OF ADVICE:

The BLACK PROJECT 'bosses' inserted this 'FAIL SAFE' programme option into place, and in case you had not realized, like rats leaving a sinking ship, jumped vessel and are no longer here!! So, what do they care, but YOU should.

There will be a release from Washington within a month, regarding Extra Terrestrial contact, which, as some of you may realize, have been going on for some considerable time.

The agency giving out this statement is not aware of the above scenario, thus the release will not be a FULL disclosure. However, we are informing you in order that you give some serious consideration as to the real driving force behind what you are doing.

REMEMBER THE LENGTHS THEY WENT TO WHEN THEY WERE RUNNING THEIR EXPERIMENTS ON THE AMERICAN PEOPLE, "UNKNOWN" TO THOSE PEOPLE, WHILE THEY CONSTRUCTED THE ATOMIC BOMB!

Think, – **until** the 21st December 2000 (actually realized in our time frame as December 21st 2001) a protective grid is and will remain in place. This grid will prevent this scenario from occurring, however, after that time - it's up to you all.

HAARP is a scientific endeavour aimed at studying the properties and behavior of the ionosphere, with particular emphasis on being able to understand and use it to enhance communications and surveillance systems for both civilian and defence purposes.

The HAARP program is committed to developing a world class ionospheric research facility consisting of:

- The ionospheric research instrument (IRI), a high power transmitter facility operating in the HF frequency range. The IRI will be used to temporarily excite a limited area of the ionosphere for scientific study.
- Diagnostic instruments that will be used to observe the physical processes that occur in the excited region.

Chapter 8

BEWARE - FALSE PROPHETS!

The information that supports the following contents of this chapter may be found in other chapters throughout this book, where we deal with the grid and timelines.

On June the 22nd 2000 the **Galactic Federation** in their ships entered, as we stated they would, through the portal opened in the protective grid, this portal was then closed. These ships will proceed to systematically carry out the task of assisting all those whom are of the Covenant in addition to those who have clearly made their free will, and wish to exit this planet prior to the commencement of the forthcoming battle. As of this point in time the battle began on December 21st 2000. Had Esau, for any reason, launched an attack prior to that date directed towards either the ships inside or any of the Galactic Federation ships which remain in close proximity on the outside of the grid, this would cause that date to be moved forward accordingly. The ships inside the grid would then be pulled out, and all remaining at that time would be here for the duration of the battle.

At the present time Esau believes that he will win that battle.

While Esau is presently preparing for the battle, he is about to implement his plan by attempting to block people from exiting, by denying them their right to free will choice.

In this section we will begin to lay out the scenario that we believe Esau is about to play out.

We stated previously that Esau believes that he can win this yet to be fought battle. Some time ago, Viking Remote viewing managed to tap into a meeting Esau had a with the Draco (Draconians), including a few other groups on the planet Pluto. This meeting was held while the battle was being fought around Jupiter. This battle was considered to be a major battle as it occurred at a point where the negative and positive grid lines crossed. However, the Galactic forces intervened at this juncture and prevented a potential threat by the negative forces from entering into the positive time line.

This plan however, began to disintegrate even before it had a chance to unfold. That plan failed when a section of their forces arrived on the 5th May 2000, only to find out that they were blocked from entering this space by the strength and presence of the BLUE GRID.

Almost all of these forces have vacated this space since, many asking for forgiveness as they departed. It was at this time you may recall that the ZETAS' asked to be freed from the inside of the grid, this was agreed upon and they were led outside the grid.

They had previously struck a deal with Esau and associates here on Terra (Earth), where they would help them develop their technology and in return would be allowed to run several biological tests on humans. They were further promised that when they were to leave this planet, they would be allowed to take with them a great number of human entities. Needless to say, it slipped their memory that they were dealing with Esau indirectly, and thus when it came time to leave they were informed they would not be allowed to take any human life forms with them.

Needless to say this plan fell on stony ground, which brings us to the plan that Esau is now about to implement.

Initially, we will explain the overall view of this plan.

Esau has several of his own ships here inside the red grid, as well as several of his own people. He will now begin to set up a situation where he will be able to materialise his equipment and try to stop the flow of those whom, by free will choice, will be exiting. In order to do this a sequence of events has been initiated, at which time Esau will bring forth a saviour to stop this flow of free will entities from departing. At this juncture Esau believes he will win this war, and feels that by letting some go before he can implement his plan, he will be able to use it as a form of leverage to execute out his new plan.

THE PLAN:

This plan is laid out in three parts:

1) DISCLOSURE...limited disclosure of longstanding dealings with alien entities.

2) CHAOS......create a situation which seems out of control, possible war OR threat of war.

3) SAVIOUR.....bring forth an alien entity to help us solve our totally out of hand world situation.

IMPLEMENTING THE PLAN

The implementation of this plan is now in full steam ahead mode.

THE GREAT DECEPTION

PROMETHEUS BOUND:

Prometheus Bound, the ancient Greek play by Aeschylus. Prometheus stole fire (knowledge) from Olympus and gave this resource to humankind, and for this act was punished by Zeus.

Our 'thought for the day' is that in this scenario a god blesses mortals, but in turn the god is punished by the chief god. What does this suggest about the nature of power in the divine order? Is it stable or unstable? A few of the gods viewed Prometheus sympathetically thus the will of Zeus may not be shared by all the gods.

We live in interesting times! This seems to be the key word banded about today, and judging by what has and is occurring, we can also add adjectives such as bizarre, weird and a feeling of concern.

Recently NASA "divulged" into the public domain (how gracious) the possibility of water on Mars;

MOC Images Suggest Recent Sources of Liquid Water on Mars

MGS MOC Releases MOC2-234 to MOC2-245, 22 June 2000

> Gullies seen on martian cliffs and crater walls in a small number of high-resolution images from the Mars Global Surveyor (MGS) Mars Orbiter Camera (MOC) suggest that liquid water has seeped onto the surface in the geologically recent past. The gully landforms are usually found on slopes facing away from mid-day sunlight, and most occur between latitudes 30° and 70° in both martian hemispheres. The relationship to sunlight and latitude may indicate that ice plays a role in protecting the liquid water from evaporation until enough pressure builds for it to be released catastrophically down a slope. The relative freshness of these features might indicate that some of them are still active today—meaning that liquid water may presently exist in some areas at depths of less than 500 meters (1640 feet) beneath the surface of Mars.
>
> The evidence for recent water activity is described in a paper by MGS MOC scientists being published in the June 30, 2000, issue of *Science.*The gullies are rare landforms that are too small to have been detected by the cameras of the Mariner and Viking spacecraft that examined the planet prior to MGS.

Taken from http://mars.jpl.nasa.gov/mgs/msss/camera/images/june2000/index.html

Don't they read their own in-house reports? Back in December 1980 Dr. Leonard Martin of Lowell Observatory published in the NASA Activities (Dec. 1980, vol. 11, number 12) findings concerning a waterspout on the surface of Mars.

In addition, what of the additional anomalies that have been uncovered over the many years, such as the Cydonia region, the 'face' and more recently the obviously 'unnatural' constructed T shape structure, termed ATREIDES FORT.

One wonders what may be revealed in the coming months. In fact we propose, judging by past experience and intensive research carried out, we know exactly what is transpiring.

In recent work we put forward the many and often ill conceived, covert operations in place around the globe. A preparation time is under way, a preparation for the intended take over of this planet by one we have called Esau and his band of followers.

Very soon this "preparation" will escalate and the 'Great Plan', or in some circles known as the 'Great Deception' will take place. The main thrust of this 'plan' is to reveal into the public domain that we have in fact been visited by extra terrestrials over time and that there has been interaction and co-operation between certain alien races and sections of our military government.

Only recently, on 12th June 2000, it was revealed by Dr Greer, founder and director of **Center for the Study of Extraterrestrial Intelligence** (CSETI) http://www.cseti.org/ that;

> The long-awaited process to disclose definitive evidence and proof regarding extraterrestrial intelligence, UFOs and covert government projects related to the subject has been funded and launched.

Indeed, this is good news! **OR IS IT?**

Before we continue, we offer a cautionary note as to what will be disclosed, as the saying goes, "Do not believe all you see or hear" Remember the three monkeys, see no evil, speak no evil, hear no evil! Take this on board as we go through the travail of the following months ahead.

Whilst we appreciate the complexities of the usual mis and dis information that is always associated with this type of subject, we did in fact offer an explanation towards an understanding of recognising the 'good over evil' as detailed in previous chapters.

We understand that this 'disclosure' will be fairly imminent, and will be broadcast simultaneously around the globe

THE DISCLOSURE PROJECT ANNOUNCEMENT June 12, 2000

Dr. Steven M. Greer announced today that the long-awaited process to disclose definitive evidence and proof regarding extraterrestrial intelligence, UFOs and covert government projects related to the subject has been funded and launched.

The Disclosure Project, an investigative reporting news organization, has been formed to spearhead and coordinate this disclosure over the next year. Key features of the project include:

A prolonged, sustained and coordinated multimedia global release of definitive evidence and the testimony of top secret witnesses about extraterrestrial intelligence, secret and compartmented government programs and advanced technologies related to energy and propulsion systems;

An integrated strategy which includes a multi-part investigative news special, compendia books and transcripts of government witness testimony, an Internet site dedicated to the disclosure evidence and on-going executive briefings for government, military, scientific, congressional, international, and corporate leaders;

A de facto international government disclosure by virtue of the large numbers of military, intelligence, government, corporate, scientific and government-related institute witnesses now ready to come forward during this integrated and sustained program;

A news conference during which top-secret witnesses disclose the full nature of their direct knowledge concerning the subject.

Authors' note. This Disclosure has now taken place in the form of a news brief in Washington DC [1](Endnotes)

Currently, over 200 top-secret witnesses have been identified in the US, Brazil, United Kingdom, Denmark, Germany, Norway, Belgium, Russia, Italy and other countries. It is important to note that during briefings at the Pentagon and with senior congressional leaders and Executive Branch staff that this civilian effort to disclose this matter has been repeatedly green-lighted and even encouraged.

The Disclosure Project and its related strategies are supported by a team of senior researchers and consultants with substantial expertise in military affairs, international affairs, national security, science, technology and related areas.

This international team working intensely on the disclosure process has been recently joined by a former senior aerospace executive - a person who has been aware since the 1970s that the Ballistic Missile Defense Program would be used to target extraterrestrial objects in space - even though there is no evidence of any credible threat from these craft.

It has been determined that the combination of accelerating global environmental degradation and the dangers associated with the weaponization of space necessitates a near-term disclosure process. The risks of doing nothing and the benefits to humanity resulting from disclosure make it clear that such an effort - long overdue by most accounts.

Taken from the CSETI website June 2000

One wonders, however, who has funded this operation?

It has come to our attention, and naturally no big surprise to many, that the Rockefeller family is involved in this promotion.

In 1999, the New York Daily News ran a story on the 85-year-old billionaire philanthropist Laurence Rockefeller who was funding a report on UFO's to be sent to President Clinton and other world leaders.

The report was entitled "The Best Available Evidence", and contained testimonies from former military officials and astronauts, contradicting denials of alien landings. The full report is entitled ***The Rockefeller - UFO Connection: An Alien Conspiracy*?**

The Rockefellers (originally Rockenfelder) immigrated in the early days to the USA, where the monopoly of the oil company, started with John D Rockefeller, and the alien agenda. In fact a book was written by John Chernow ***Titan*** **-** *The Life of John D. Rockefeller, Sr,* 1998 Random House; ISBN: 0679438084, where it goes into great detail concerning the strategies of John D. Rockefeller and the Standard Oil Company 1863-1911.

Upon further research, it was discovered that the Standard Oil Company was the same as that as I.G. Farben, Hitler's Chemical plant which operated the concentration camp at Auschwitz. (D.Icke 1999 *The Biggest Secret* 254-256) Standard oil is better known today as EXXON

For those of you who may be interested in the genealogy of European families and logos, the symbol of the double XX is that of the House of Lorraine, often associated with literally a double cross or the concept of manipulation!

The Rockefellers, Rothschilds, Harrimans and Bushs were all behind the Nazi race purity party programme, headed by one Dr. Ernst Rudin from the Kaiser Wilhelm Institute for Genealogy and Demography in Berlin - research funded by the Rockefellers. (Webster Griffin Tarpley and Anton Chaitkin ***George Bush, The Unauthorised Biography*** - Executive Intelligence review, Washington DC 1992 p42)

The Rockefellers are naturally, a well-known family worldwide with a number of well-established institutes named after the family, such as Rockefeller University, New York In addition, they have a number of websites with funding opportunities such as the Rockefeller Brothers' fund and the family fund site at http://www.rffund.org/ all supporting a wide range of environmental, political participation and reform.

It is interesting to note that a gold statue of Prometheus stands in the Rockefeller center in New York. Bearing the initial 'thought of the day' in our opening section, does this reflect the theme of the fallen angels giving forbidden secrets to humanity?

The family under J.D Rockefeller, Nelson Rockefeller, Winthrop Rockefeller, Laurence Rockefeller, while the most notorious, in middle of 20th C, David Rockefeller head of Chase Bank and behind all presidents, controlling the money, media and politics, have built up one of the most influential sections of our communities.

> Television commentator Bill Moyers found out during a fifteen-day, globe spanning trip in the company of David Rockefeller that 'just about a dozen or fifteen individuals made day-to-day decisions that regulated the flow of capital and goods throughout the entire world.'"

This was a quote from Father Malachi Martin, in his ***The Keys of This Blood***

Father Malachi was a profuse writer and Vatican advisor, and unfortunately for all, died mid 1999. It was Father Malachi who brought to the attention of the public, the interest of the Vatican in astronomical observatories and their real purpose in stargazing.

For some time the reptilian brotherhood has been preparing the prevention of humanity in making the consciousness shift - attacks on physical level, drugs, food additives, fluoride in water etc. And it is the family that we have mentioned above that actually own over 60% of the pharmaceutical industry in the USA.

It is recognised that past UFO research has been funded by Laurence Rockefeller, and that many are associating this forthcoming 'exposure' with a research project known as "Project Bluebeam" [22] **Project Blue Beam (from Contact, April, 1996)**

Set to take place this year is the worldwide, simultaneous, Christian Crusade, Billy Graham, all languages audio and video revealing of our "God" in the clouds. This is also known as Operation Blue Beam. It got shut down last year but, to fulfill the expectations of the "people" for a Rapture and Second Coming, etc., there MUST be a visualization in every part of the world - to allow for the ONE WORLD RELIGION.

The "revelation" herein is to stop, look and listen before you jump off the cliff into the arms of an illusion. The "operation" is easy to EXPLAIN and REVEAL technologically.

Holography, a technique for obtaining 3-dimensional images, involves an interference pattern between 2 sets of single-wavelength light waves. A single laser beam is split into two parts, one to illuminate the object, the second part (called the reference beam) is reflected by a mirror. The two beams meet forming an interference pattern, which is developed to form a hologram.

This hologram can be used to re-create a 3-dimensional image of the original object. According to a comprehensive report by Canadian investigative journalist, Serge Monast, entitled Project Blue Beam, this holographic technology used in conjunction with hi-tech, mind-control weapons utilizing the effects of radio-frequency waves upon the brain, is capable of projecting an image in the sky and causing collective thought to convince people they are seeing an alien invasion or the second coming of Christ.

The Blue Beam Project will pretend to be the universal fulfillment of the prophecies of old. In principle, it will make use of the sky as a movie screen space-based, laser-generating satellites project simultaneous images to every part of the planet, in every language, in every dialect according to region. Specifically, the show will consist of laser projections of multiple holographic images to different parts of the world, each receiving different images according to predominating regional/national religious faiths. No area will be excluded. With computer animation and sound effects appearing to come from the depths of space, astonished followers of the various creeds will witness their own returned Messiah in spectacularly convincing lifelike realness.

The various saviors will then merge into one after "correct" explanations of the mysteries, prophecies, and revelations will have been disclosed. This event will occur at a time of great political and general tumult. [Seems like surely this is a good time to meet those requirements.]

The Blue Beam Project, developed by NASA scientists, consists of 4 parts, 4 steps toward the implementation of the NWO religion with the Antichrist at its head.

First, earthquakes at certain precise locations of the planet will result in new archaeological discoveries, revealing that all religions' basic doctrines have been misunderstood and misinterpreted.

Second, a gigantic Space Show with laser projections of multiple 3-dimensional holographic images worldwide, will appear to be the image of God, speaking in all languages.

This project utilises holograms and other high tech gadgetry, in order to convince the population that we are being invaded by Extra Terrestrials.

So, it is with this background and under these circumstances that events and disclosures abound over the forthcoming months. Bear in mind that while motives driving the disclosure are a worthy undertaking, one must keep an open mind as to the slant which it will take in an attempt to direct us towards. Dr.Greer and others involved with **CSETI** have tried for years to bring this situation out into the public arena. How convenient is it RIGHT NOW that this is about to be bankrolled by one of the largest power merchants on the planet? Part I of the plan is thus implemented.

As we noted above, this disclosure has now taken place, on May 9th

2001, and "coincidentally" is practically nearly 12 months since the original given date of this disclosure in May/June 2000. For further information on this subject please refer to chapter 15.

PHASE 2

In order for any of the plan to be carried further, a crisis situation must then come to light. (Do you get the feeling that we have a ***WAG THE DOG*** show about to unfold?). The difference here, however, is that there is much more at stake, and in order to further move the plan ahead, how many crises will they create.

Sit back for one minute and take stock of the more volatile situations that were currently underway at the time of writing;

Here are but a few potential scenarios, if not all.

1) Talks between North and South Korea.
http://www.cnn.com/2000/ASIANOW/east/06/18/korea.us.ap/index.html

2) The Iraq situation.
http://news.bbc.co.uk/hi/english/world/americas/newsid_809000/809168.stm

3) Middle East - Israel Dispute
http://news.bbc.co.uk/hi/english/world/middle_east/newsid_808000/808333.stm

4) U.S.A.-Russian, China disagreement - Reference Executive Order issued on **JUNE 22nd** re: Nuclear situation-[33] http://www2.50megs.com/giza2012/Exec.html

EARTHQUAKE EVENTS:

Whilst preparing this research it has come to our attention there will be an event that will be part of the aforementioned scenario. This event is the occurrence of three earthquakes, taking place over a short span of time.

The locations in which they occur are as follows:

- **Murmansk, Russia....approx. N 68° latitude, E 35° longitude**

- **Durban, South Africa…approx. S 30° latitude, E 30° longitude**
- **Las Vegas, Nevada, USA…approx. N 37° latitude, W 117° longitude**

Upon looking at a map of the world, the three above locations form a triangle, at the centre of which are the Azores Islands.

Extrapolating a line from the Azores through the centre of the earth, the resulting exit point is offshore Sydney, NSW, Australia. Both visionaries and the eastern states **Aboriginal prophecies speak of a tsunami that wipes the shoreline clean at Sydney.**

The three earthquakes will set in motion a series of events that will lead to the appearance of the false prophet.

The same three areas were hit by earthquake in 1994/95, but there was no loss of life. A search on the NEIC database (http://gldss7.cr.usgs.gov/neis/bulletin/bulletin.html) shows this to be true. On this occasion however, there will be loss of life.

This is a warning of things to come.

In addition, we have also noted that some unusual earthquake and volcanic events have occurred during the last week of June 2000. In particular, the eruption of a volcano on the Japanese island of Miyakejima.

On June 7th 2001, we stated on the Mitch Battros "Earth changes TV" radio show, that there would be a significant star alignment a few days later, that is, on the June 10th. Below is the information we received immediately after June 10th.

PHOENIX FIVE EARTH CHANGE BULLETINS
June 10, 2001 by MW Mandeville
ITEM(S): Such A Day: 6:20 am shake in Vancouver and energy through all systems is suddenly way up

Martin, that is not all.

As long predicted by the principles of Vortex Tectonics in these bulletins and at the Phoenix websites (MichaelMandeville.com) the energy flowing through all systems is suddenly surging up in response to the Inner Planet Alignment. Sunspots have increased suddenly in the space of 48 hours from 150 to 260.

The Earth's Magnetic field is reacting and showing that we are sailing in an increasingly stormy solar ocean with the chop factor up significantly.

The potential of sudden X class or M class solar flares in radically higher and warning alarms are beginning to flow out through many channnels. And more,during the last 48 hours, seismic activity in California (all low level quakes) is suddenly much higher than normal and the same is true in the PNW, where you felt a quake in Vancouver and where the roots of Washington's Olympic Peninsula shook at 5.0.

This all flows after two weeks of unusually high levels of seismic activity deep deep in the spreading rift of the South Pacific. The numbers of 5.0 to 7.0 quakes in this one area of the eastern part of the Fiji-Papua Tectonic Arc is outclassing all other areas.

I think it is pretty safe to describe the situation in the South Pacific as currenlty the prime generator of seismic stress in the Earth's crust. Sitting virtually on the Equator, this area of the Earth is, in a Cayce word, upheaving and splitting apart, literally forcning the other plates to crack, creak, quake, shake, and move. I think we can add it as a "prime suspect" to the principles and findings of Vortex Tectonics that all of this generator activity in the South Pacific is reflected around the entire Pacific Rim, resulting in "answering" earthquakes within a few hours to a few days. I don't really want to do anymore graphs without some formal research support but this stuff is awfully damn compelling. I may just have some time after I get more rust scrapped off my truck.

This increased tectonic motion along the Pacific equator is most likely directly connected to the demonstrable increase in the rate of shifting of the locus of Chandler's Wobble (the average location of the pole) since 1998.

> It is forcing the South Pacific to ride up higher onto the Equatorial Rim.
>
> Speculation: this could be a sign that the Earth is beginning to escape its hydraulic clutch breaks at the crust/mantle boundary zone by an escalation of the rate of deformation in the equatorial shape. This would release the pressure, increase the North-South motion of the

wobble, thus allowing the crust to wobble even futher "off the old center", and naturally, increase the total instability in the system.

At this point, the Cosa Nostra predictions are tectonically speaking highly credible and we can see the trigger being cocked day by day in the Fiji-Vanuatu Islands. It is stiff ratcheting up to another click.

WATCH JUNE 21-25. Elaine Phibbs-Earl, who was remarkable in naming March 23 as the end of "peace", tunes to June 25 as the timing which will bring some of the Cosa Nostra troubles. From sygygy principles alone, I would predict June 21-23 as the most dangerious period, but the Lunar Perigee on June 23 places June 25 within the syzygy window.

With reference to the above information, one should note that the time difference between British Vancouver and Cairo is 10 hours in advance. It is also of note that the constellation of Delphinus constituted part of that alignment.

PHASE 3

By this time we will have discovered the technologies that have been developed over the last 60 years, not to mention the source of these technologies, having been supplied to us by alien beings. In addition, these same trustworthy aliens have been monitoring our progress and now wish to help solve our global problems - ENTER ESAU'S ARMY.... come to protect the people of Earth from one another. To unsuspecting trusting minds, the Prophet/Messiah has come to bring PEACE to the people of the world...

This illusion will be short lived, however, as we KNOW why Esau has part of his army here. They will be here in an attempt to prevent people from exercising their right to free will choice. Do not put it past Esau to begin to physically round up members of the public and place them in the concentration camps he has in place now worldwide.

When we initially released this particular segment of information, we had been aware of a number of attempts to create and instigate certain situations. However, we are unable to release our source on this incident at the present time, but we do know "they" received the information and the following message that we posted on our Ambilac website:

> **Now for a clear message to those who believe that they are in control of the EVENTS taking place within this space.**
>
> **Are we a little confused boys as to why things never happened the way you played them out last week??**
>
> ***"THERE IS ALWAYS "ONE" WHO IS STRONGER"***
>
> **Keep in mind boys that the "SPIRIT" is stronger than those implants, and all you have to do to get out from under ESAU'S control is to declare your "FREE WILL CHOICE" to do so.**

Third, electronic telepathy involving ELF, VLF, and LF waves will reach people by the inside of their brains, interlacing and interweaving with the natural thinking to form diffused artificial thought and making each one to believe that God is speaking to them from within their own souls.

Fourth, electronic universal "supernatural" manifestation designed to deceive will create the following illusion: make mankind believe an alien invasion is imminent, make Christians believe a rapture is imminent and the aliens have come to rescue them, and convince all that global satanic supernatural forces and manifestations penetrating worldwide - able to travel through optical fiber, coaxial cable, electric and phone lines - are everywhere and inescapable.

(Endnotes)

[1] From an eye witness account of the Disclosure news brief Washington DC (from CSETI site)

> Lance, I was there live in the room as part of the Disclosure Project Staff, and it was in fact a "God Given Privilege" to be a part of that historic event.
>
> As you saw from the note from ConnectLive, we were jammed on the Livecast. Please do not blame ConnectLive, they were wonderful at trying to counter the RF and other sophisticated interference that was going on.

They have NEVER had this happen to them before, which shows the lengths some people will go to try to prevent what we were doing.

Just a few statistics we got that evening, many direct from ConnectLive.

1. At the start of the National Press Club conference, there were 250,000 people logged on ready to watch. That is one quarter of a million people! But that did not crash their system. We were very definitely jammed, probably trying to discourage various media who would not be persistent enough to come back later.

2. As of 9:00 PM that night, we still had 13,000 people on simultaneously viewing the archive. They had 10 T-1 lines tied up just broadcasting us.

3. The previous record was 20,000 people on line live, and that was to watch the Monica Lewinski thing. We were over an order of magnitude larger than that, by far the largest livecast in their history.

4. Video's of the Press Conference are now available, but I don't have any statistics on sales yet. I do know that people are buying the video in droves and doing an end run on the press to see the story themselves.

This is another way to vote, by the way. If Connect Live sells 100,000 units or so, that would be powerful evidence to our representatives in the House and Senate that THE PEOPLE WANT TO KNOW!

5. We had the largest crowd in the audience (238 signed in) of any event with the possible exception of when President Regan spoke there as a sitting President.

6. We had more cameras in the room than anyone could remember. I did not count, but guessing 35-40 of the Big professional cameras. I have photos, and perhaps when they get the film developed, you can see some on the web.

7. Media experts say that already over ONE BILLION people worldwide have seen or heard the story!

8.The story has been largely ignored in this country, but stay tuned! Popular Mechanics is going to follow up last months fine article with a major article on us, The Wall Street Journal is working on a major piece, and we have been all over talk radio. Ted Koppel is looking into the story. But that is small potatoes. The story is sweeping through France, Spain, Europe, Asia, Australia, and Latin America. The Tsunami is building, and it will travel all the way around the world and will come back to swamp our complacent (or complicit) press in their own lies.

9. We had an ABC journalist stand up and tell the audience that he has had similar stories spiked in the past, and he was mad about it. He was probably jeopardizing his job to say that. Hopefully you could hear that on the coverage.

10. We presented 22 witnesses out of 100 already video archived, and over 400 on file. This is like 5% of what we had at that time. Many more witnesses were coming forward this week and contacting us, so those numbers will undoubtedly swell. This is just the tip of the iceberg. Many of our Most Explosive witnesses are waiting to be sworn in before congress and given protection. Taking nothing away from the very high caliber of the first round of witnesses, the best is definitely yet to come. This is getting a bit long, so I will write another note on how much fun we had wandering the halls of the House and Senate office buildings on Thursday and Friday.
Don Daniels

[1] **Project Blue Beam (from Contact, April, 1996)**

Set to take place this year is the worldwide, simultaneous, Christian Crusade, Billy Graham, all languages audio and video revealing of our "God" in the clouds. This is also known as Operation Blue Beam. It got shut down last year but, to fulfill the expectations of the "people" for a Rapture and Second Coming, etc., there MUST be a visualization in every part of the world - to allow for the ONE WORLD RELIGION.

The "revelation" herein is to stop, look and listen before you jump off the cliff into the arms of an illusion. The "operation" is easy to EXPLAIN and REVEAL technologically.

Holography, a technique for obtaining 3-dimensional images, involves an interference pattern between 2 sets of single-wavelength light waves. A single laser beam is split into two parts, one to illuminate the object, the second part (called the reference beam) is reflected by a mirror. The two beams meet forming an interference pattern, which is developed to form a hologram. This hologram can be used to re-create a 3-dimensional image of the original object. According to a comprehensive report by Canadian investigative journalist, Serge Monast, entitled Project Blue Beam, this holographic technology used in conjunction with hi-tech, mind-control weapons utilizing the effects of radio-frequency waves upon the brain, is capable of projecting an image in the sky and causing collective thought to convince people they are seeing an alien invasion or the second coming of Christ.

The Blue Beam Project will pretend to be the universal fulfillment of the prophecies of old. In principle, it will make use of the sky as a movie screen space-based, laser-generating satellites project simultaneous images to every part of the planet, in every language, in every dialect according to region. Specifically, the show will consist of laser projections of multiple holographic images to different parts of the world, each receiving different images according to predominating regional/national religious faiths. No area will be excluded. With computer animation and sound effects appearing to come from the depths of space, astonished followers of the various creeds will witness their own returned Messiah in spectacularly convincing lifelike realness. The various saviors will then merge into one after "correct" explanations of the mysteries, prophecies, and revelations will have been disclosed. This event will occur at a time of great political and general tumult. [Seems like surely this is a good time to meet those requirements.]

The Blue Beam Project, developed by NASA scientists, consists of 4 parts, 4 steps toward the implementation of the NWO religion with the Antichrist at its head.

First, earthquakes at certain precise locations of the planet will result in new archaeological discoveries, revealing that all religions' basic doctrines have been misunderstood and misinterpreted.

Second, a gigantic Space Show with laser projections of multiple 3-dimensional holographic images worldwide, will appear to be the image of God, speaking in all languages.

Third, electronic telepathy involving ELF, VLF, and LF waves will reach people by the inside of their brains, interlacing and interweaving with the natural thinking to form diffused artificial thought and making each one to believe that God is speaking to them from within their own souls.

Fourth, electronic universal "supernatural" manifestation designed to deceive will create the following illusion: make mankind believe an alien invasion is imminent, make Christians believe a rapture is imminent and the aliens have come to rescue them, and convince all that global satanic supernatural forces and manifestations penetrating worldwide - able to travel through optical fiber, coaxial cable, electric and phone lines - are everywhere and inescapable.

[1] EX Order THE WHITE HOUSE Office of the Press Secretary (Phoenix, Arizona)

Immediate Release June 22, 2000 EXECUTIVE ORDER - - - - - - - BLOCKING PROPERTY OF THE GOVERNMENT OF THE RUSSIAN FEDERATION RELATING TO THE DISPOSITION OF HIGHLY ENRICHED URANIUM EXTRACTED FROM NUCLEAR WEAPONS

By the authority vested in me as President by the Constitution and the laws of the United States of America, including the International Emergency Economic Powers Act (IEEPA) (50 U.S.C. 1701 et seq.), the National Emergencies Act (50 U.S.C. 1601 et seq.), and section 301 of title 3, United States Code. I, WILLIAM J. CLINTON, President of the United States of

America, in view of the policies underlying Executive Order 12938 of November 14, 1994, and Executive Order 13085 of May 26, 1998, find that the risk of nuclear proliferation created by the accumulation of a large volume of weapons-usable fissile material in the territory of the Russian Federation constitutes an unusual and extraordinary threat to the national security and foreign policy of the United States, and hereby declare a national emergency to deal with that threat. I hereby order: Section 1. A major national security goal of the United States is to ensure that fissile material removed from Russian nuclear weapons pursuant to various arms control and disarmament agreements is dedicated to peaceful uses, subject to transparency measures, and protected from diversion to activities of proliferation concern. As reflected in Executive Order 13085, the full implementation of the Agreement Between the Government of the United States of America and the Government of the Russian Federation Concerning the Disposition of Highly Enriched Uranium Extracted from Nuclear Weapons, dated February 18, 1993, and related contracts and agreements (collectively, the "HEU Agreements") is essential to the attainment of this goal. The HEU Agreements provide for the conversion of approximately 500 metric tons of highly enriched uranium contained in Russian nuclear weapons into low-enriched uranium for use as fuel in commercial nuclear reactors. In furtherance of our national security goals, all heads of departments and agencies of the United States Government shall continue to take all appropriate measures within their authority to further the full implementation of the HEU Agreements. Sec. 2. Government of the Russian Federation assets directly related to the implementation of the HEU Agreements currently may be subject to attachment, judgment, decree, lien, execution, garnishment, or other judicial process, thereby jeopardizing the full implementation of the HEU Agreements to the detriment of U.S. foreign policy. In order to ensure the preservation and proper and complete transfer to the Government of the Russian Federation of all payments due to it under the HEU Agreements, and except to the extent provided in regulations, orders, directives, or licenses that may hereafter be issued pursuant to this order, all property and interests in property of the Government of the Russian Federation directly related to the implementation of the HEU Agreements that are in the United States, that hereafter come within the United States, or that are or hereafter come within the possession or control of United States persons, including their overseas branches, are hereby blocked and may not be transferred, paid, exported, withdrawn, or otherwise dealt in. Unless licensed or authorized pursuant to this order, any attachment, judgment, decree, lien, execution, garnishment, or other judicial process is null and void with respect to any property or interest in property blocked pursuant to this order. Sec. 3. For the purposes of this order: (a) The term "person" means an individual or entity; (b) The term "entity" means a partnership, association, trust, joint venture, corporation, or other organization; (c) The term "United States person" means any United States citizen; permanent resident alien; juridical person organized under the laws of the United States or any jurisdiction within the United States, including foreign branches; or any person in the United States; and (d) The term "Government of the Russian Federation" means the Government of the Russian Federation, any political subdivision, agency, or instrumentality thereof, and any person owned or controlled by, or acting for or on behalf of, the Government of the Russian Federation. Sec. 4. (a) The Secretary of the Treasury, in consultation with the Secretary of State, the Secretary of Energy, and, as appropriate, other agencies, is hereby authorized to take such actions, including the promulgation of rules and regulations, and to employ all powers granted to me by IEEPA, as may be necessary to carry out the purposes of this order. The Secretary of the Treasury may redelegate any of these functions to other officers and agencies of the United States Government. All agencies of the United States Government are hereby directed to take all appropriate measures within their statutory authority to carry out the provisions of this order. (b) Nothing contained in this order shall relieve a person from any requirement to obtain a license or other authorization from any department or agency of the United States Government in compliance with applicable laws and regulations subject to the jurisdiction of the department or agency. Sec. 5. This order is not intended to create, nor does it create, any right, benefit, or privilege, substantive or procedural, enforceable at law by a party against the United States, its agencies, officers, or any other person. Sec. 6. (a) This order is effective at 12:01 a.m. eastern daylight time on June 22, 2000. (b) This order shall be transmitted to the Congress and published in the Federal Register.

WILLIAM J. CLINTON THE WHITE HOUSE, June 21, 2000.

Chapter 9
POLITICAL INTRIGUE

Recent and increasing earthquake events, and indeed further events due shortly, may possibly induce a serious knee jerk reaction in certain military circles. These natural earthquake events (we are also aware of the technology that may possibly induce such events) will be centered on, or around well-known facilities relating to nuclear storage and armament depots. These events, at this point in time, are considered to be purely natural events and not contrived as others have been to date, thus we advise caution to those not to instigate any knee jerk counter defensive measures.

As far as we can ascertain the location of these forthcoming events are located at Murmansk Russian Federation, Las Vegas Nevada, USA and Durban South Africa.

Whilst all locations are important in this scenario, one particular location will bring the hairs to stand to erection as it were, and that is the Berents Sea area, (Arctic circle) and notably the city of **MURMANSK**, in the Russian federation.

It is interesting to note that not too far distant from Murmansk, on the coast of the White Sea, can be located the town with the intriguing name of **ARCHANGEL**.

Apart from a very recent earthquake at this location, there have been several previous quakes of note, one in January 1994, a 2.7, and one in August 1999, a 4.6.

(US Geological survey earthquake database circle centre, point Latitude 68.000N. Longitude 35.000E.)

Murmansk is situated on the Kola Gulf of the Berents Sea, and is the terminus of the Northeast Passage. In 1989 its population was 468.000. As a major ice-free port it is a base for naval and fishing vessels and the world's largest city North of the Arctic Circle. Allied forces occupied Murmansk in 1918-20, and during World War 2 was a major supply base and port for Anglo-American convoys.

In September 1999, the Murmansk Radwaste facility (liquid radwaste) was completed but not actually commissioned, and at the time of writing no details were forthcoming as to the actual date of commissioning, however a brief was written by the United Institute of Geology, Geophysics and Mineralogy, Russian Academy of Sciences, Siberian Branch:

More importantly, a report on the sightings.com news web site, (http://www.sightings.com/politics4/rusec.htm) revealed that a Russian secret chemical weapons plant had been operating for 15 years, just east of Murmansk. This facility is producing and storing deadly chemical weapons.

> An explosion at this plant could lead to an environmental catastrophe in the Nordic region. There would be damage up to 150 miles away.

Quote from Verdens Gang, Norwegian daily.

In addition, on Monday 14th August 2000, there was a serious accident involving a Russian submarine (The Kursk) in this very area of the Barents sea, of which the BBC news report is included below;

> BBC on-line news
> Monday, 14 August, 2000, 11:00 GMT 12:00 UK
> **Russian nuclear sub crippled**
>
> A Russian nuclear submarine has broken down during exercises in the Barents Sea, and is trapped on the sea floor with a crew of more than 100 on board.
>
> The Oscar-II class submarine was not carrying any nuclear weapons and there was no immediate danger, said the head of the Russian navy's press service, Igor Dygalo.

He said no radiation leaks had been reported.

Rescue ships are rushing to the area to assist the crippled submarine, named Kursk, which is in radio contact with surface vessels.

Oscar-II class submarines usually carry a crew of 107 personnel, including 52 officers.

A correspondent for Russian NTV television reported the crew had been forced to ground the Kursk after problems with its torpedo tubes flooded the vessel's front section.

There was much controversy surrounding this "accident", and one may wonder if indeed the submarine it was actually hit by a beam via HAARP. If one reads the below BBC news item of a few months later, one could seriously question the whole scenario.

Tuesday, 21 November, 2000, 11:22 GMT (BBC On line news)

New evidence in Kursk disaster puzzle

A top Russian navy commander says that SOS signals from a foreign submarine were recorded when its own nuclear submarine, the Kursk, was sinking in August.

One hundred and eighteen sailors died in the accident on the Barents Sea bed.

The commander of the Russian Northern Fleet, Vyacheslav Popov, is reported to have told a newspaper Nezavisimaya Gazeta that the signals had been picked up by a navy cruiser Peter the Great.

Laboratory analysis showed the signals had been emitted by a foreign submarine near the Kursk, which he said added to evidence that the disaster had been caused by a collision. But he added that the real reason behind the accident would only become clear once the submarine had been raised.

It is therefore imperative that we consider the plausibility of such an event being caused by a natural event, and not to be confused with a contrived earthquake. Therefore, under the prevailing circumstances, the possibility of a military retaliation cannot be ruled out, the affect of which would result in a scenario of nightmarish proportions.

THIS MUST NOT UNDER ANY CIRCUMSTANCES BE ALLOWED TO PLAY OUT.

LAS VEGAS, USA

We have also discovered that northeast of Las Vegas Nevada, is the Nevada Test Site, surrounded on three sides by Nellis Air Force Base, which is adjoined to the famous Area 51. This area was used for nuclear weapons testing from 1951 to 1992, using above ground and later underground tests. Nellis AFB is known also for its development of air vehicles, some notable examples being the Stealth Fighter and the Aurora.

Linked in UFO lore to Dulce, there has been purported alien interaction at Nellis AFB and Area 51. Although officially unconfirmed, there appears to also be a nuclear storage facility at Nellis.

To the north of Nevada Test Site and Nellis AFB is Yucca Mountain, which is under investigation to become the National Nuclear Repository (read "dump") of the US. Although not officially active as yet, everything seems to be in place to become active at any time. However, there are a few problems with this, one being the possibility of water problems coming from under the site. See Dublyansky's comment below:

> Comments on Draft Environment Impact Statement on Yucca Mountain, DOE/EIS-0250D
>
> Prepared by Dr. Yuri V. Dublyansky
>
> *Draft EIS does not analyze the potential impact of inundation of the repository zone by upwelling water*

Draft EIS acknowledges that inundation of the repository zone by upwelling water, if happens, would have great impact on the long-term repository performance. (1) The possibility of such inundations was suggested by a number of scientists (Szymanski, 1989; Hill et al., 1995). Draft EIS explicitly states, however, that "*DOE does not agree with the inundation scenario*" (p. 5-15). This dismissal heavily rests on the findings of the 1992 NAS/NRC panel (National Research Council, 1992). The latter document is outdated, because much new data have became available since 1992. Below we summarize some of this evidence.

> Fluid inclusion evidence
>
> By rejecting "inundation scenario", DOE rejects new scientific information indicating the presence of waters with elevated temperature in what is now Yucca Mountain unsaturated zone in the past, obtained by studies of fluid inclusion in secondary minerals. (2) "Justification" of this rejection is given on pp. 3-49 - 3-50 of the Draft EIS as follows: "*DOE, given the opportunity to review a preliminary version of the report, arranged for review by a group of independent experts, including U.S. Geological Survey personnel and a university expert. This review group did not concur with the conclusion in the report by Dublyansky (1998 all)...*"

In addition there was a highly revealing and timely article on this very same subject in the weekly UK science magazine NEW SCIENTIST

The time travelling mountain. Larry O'Hanlon.
Issue dated 7th July 2000.

Las Vegas, a city of 450,000 people, is only 65 miles away from Nellis AFB, and apart from the nuclear storehouse, we have the old test facility where there is radioactivity in the ground and surrounding flora, and a possible nuclear repository. While at the time of writing, the US wishes to set up an antimissile defence programme at this location.

Meetings have been held in Russia regarding this, with Russia not supporting the US to undertake such a move, thus there is much tension at present between the US and Russia, the two old so called "Cold War enemies".

Both countries have Electromagnetic Pulse weapons technology, which can be used to create earthquakes, amongst other effects. There is a very real threat that an earthquake-induced nuclear or bio-chemical explosion could cause the present tension to result in the pressing of the nuclear missile buttons in either country.

NOTE:

In the course of our research we receive a substantial amount of mail, however, in most cases it is electronically generated, and as is most often the case generally concerns the material that we had previously posted to one of our web-sites.

The mail below however, was received through the normal postal service, which is in itself is strange due to the fact we had NO postal address listed (least of all the address to which it was sent). However, and importantly, it is SOME of the content that proved most interesting.

We have taken the **LIBERTY** of including s few pertinent sections of this mail, which we include below.

It is accepted that some of the material we were actively researching at the time we received this mail is from a somewhat different viewpoint, however, the timing of this mail, in addition to its content, did cause us to take a closer look at the areas of research material we were accessing.

First, we include **the letter** we received (edited for brevity).

There are those who would like to come forward but who would jeopardize their safety and that of their friends and families if they did.

> It has come to my attention that the New World Order has begun implementing its master plan, which calls for limited controlled nuclear warfare. Their goal is to decrease human population levels and enslave the survivors of the holocaust (a) to control genetic development on the planet, and (b) through using the information gained through the Genome Project, bring about immortality for their members. In order to accomplish their immortality, they need a stock of DNA, which will be harvested from the slaves.

By initiating limited nuclear warfare, they can then approach the Universal Council and make attempts to enter into the Universal Federation of the Planets and Star Systems. Contact with the Universal Federation depends on the outcome of the above actions. But in fact all aspects and variables of this New World Order plan, calling for limited nuclear war, are controlled.

Having said that, here is a series of the events leading up to the present situation.

- All world leaders attended meetings, at which they all agreed to the following:

- At the conclusion of each meeting they would submit to brainwashing techniques, thereby removing all memory of the events taking place at the meeting. These memories would be re-instilled at the beginning of subsequent meetings and afterwards "removed." This practice ensures there will be no leakage of the business conducted at these meetings, while at the same time maintaining individual input from these world leaders.

These meetings were attended by alien representatives.

- During these meetings plans to bring about a one-world government were developed and agreed upon. The implementation of this one-world government was also established.

- Preparations have been under way for many years. With the help of alien races, we now have several interconnected underground cities which will house those that the New World Order has tagged.

- The military wing of the New World Order is controlled on most levels through the use of implants and other forms of mind control. The forces that make up this wing of the New World Order come from all countries.

I'm sure you're aware of most of this information, but what I wish to bring to your attention is the "Other" plan.

A group within and above the New World Order-those in control of the technology and financing-have initiated their own agenda. The following events will occur:

- Two earthquakes, one at Murmansk and the other near Las Vegas, Nevada. These earthquakes are being created, using advanced technology which allows for the control of the forces governing matter in a given time and space.
- In both locations are nuclear waste storage sites : the one in Murmansk is a Russian storage facility; the one in Nevada, Yucca Mountain, is a US storage facility.
- It is calculated there will be a nuclear response. If there is no response from the Russian side, they will create a situation by launching a limited attack on itself. That will call for a retaliatory attack on Russia. Either way, a limited nuclear strategic exchange will occur, thus reducing world population in a controlled manner.

I've written this information in longhand as I am aware that all your electronic correspondence is monitored, and I have posted this. Please put this material forward and also destroy all postmarks, etc. that may be traceable.

Authors note:

We must emphasize that all material, electronic and otherwise that has passed through our hands in our research has been disabled, so as not to be traced from its source, or point of origin.

We believe that the above scenario is not a valid one. However we have included it in this chapter as we have no way to disprove it. If such a plan did exist, the rogue element would now, with the airing of this information have placed themselves at high risk.

Secondly, if this scenario were to occur this would negate the possibility of any ET intervention or association from occurring and most likely would be terminated. This would not be in their best interest for this to unfold.

To assist towards a deeper understanding as to the scenario presently about to unfold, we submit edited sections from the research of Dr. Richard Boylan, behavioral scientist. Dr Boylan is well known in his field and over several years has had the unique opportunity of interviewing one Dr. Michael Wolf, (recently deceased) a National Security Agency (NSA) advisor.

In the interests of public knowledge and preparation, we submit a number of extracts below, complete with the URL references in order that one may acquire the complete take on the situation.

Our first extract is from an article in 1999 entitled, Aviary, *Good Bye: Hello. Acclimation Program*

The extract is from an interview on the division within the government on the UFO/ET disclosure.

The full article can be read at http://www.jps.net/drboylan/aclmatn.htm

> 1.PI-40 is the name which UFO/ET researcher Steven Greer, MD was informed was the code name for the governmental unit in charge of UFO/ET matters. This group has been in operation since 1947, when it was called MJ-12. It was originally organized by President Harry Truman to deal with the recovery of two discs and their extraterrestrial occupants in the New Mexico desert around July, 1947. Information I have received from Dr. Michael Wolf, scientific consultant to this group, has provided further detail. This group operates as an unacknowledged subcommittee of the National Security Council. Its current name is Special Studies Group. I believe that this is the proper title for this group, and thus will not use the PI-40 designation.
>
> Another informant of mine, "Jesse", with ties to the National Security Agency, tells me that the group's membership now numbers 33. Reportedly, Henry Kissinger is MJ-1, the head of the NSC Special Studies Group. Edward Teller, the father of the H-bomb, is another member. Dr. Wolf makes a third.
>
> 4.The Aviary group members who have been active in the past in openly disseminating UFO information have been Jacques Vallee, Bruce Maccabee (Navy laser weapons researcher), and Cmdr. Scott Jones, USN (ret.), (who organized a Cosmic Cultures Conference in Washington.) Some of the better-known Aviary members who have more recently dissembled any involvement in UFO matters include

Ron Pandolfi and Col. John Alexander. The more important split is occurring on the National Security Council's Special Studies Group (SSG). Progressives on the SSG have placed into operation an Acclimation Program, designed to prepare and condition the public for open acknowledgment of extraterrestrial contact and presence. This Acclimation Program has resulted in significant disclosures of UFO/ET information by highly-placed governmental officers. Col. Steve Wilson, USAF (ret.) revealed that he was in charge of Project Pounce, the unit tasked to retrieve downed UFOs and prevent civilian access to them. He also revealed the designations and manufacturers of U.S. antigravity craft. Col. Donald Ware, USAF (ret.) leaked the very existence of the Acclimation Program, and also that the B-2 bomber has an antigravity system aboard.

Continued:

The Aviary is not a governmental entity and is not running the government's UFO gradual- disclosure program. They are not responsible for the above-cited leaks, which are instead coming from NSC's Special Studies Group's Acclimation Program policy. The Aviary appears now to have been an ad-hoc group of UFO information exchangers, some of whose members were inclined to publicly share that information, and some of whose members continue to fight public disclosure. As the official government's Public Acclimation Program moves into high gear, the Aviary becomes increasingly irrelevant.

6. Among those in the Aviary who have publicly shared UFO information, there may not be a consensus about the nature of extraterrestrial visitors. Scott Jones appears to view the ETs as benign. Jacques Vallee takes a more scientifically-detached view, as does Bruce Maccabee.

7. Some of those on the shadowy side of the Aviary group work for private corporations which have made huge sums of money, developing and manufacturing military and civilian

> products derived ultimately from ET technology. Extreme security classification on ET technology has created a monopoly for those insider corporations with connections to those who possess ET technology information.
>
> 8. Is the struggle over UFO disclosure a theological war? Not the way that question suggests, but Dr. Steven Greer, Director of the Center for the Study of Extraterrestrial Intelligence, has talked with high officials in government. He has learned that a number of right-wing extremists in the Cabal are also fundamentalist religious extremists, who demonize the extraterrestrial visitors, and may view their private war as a "crusade".
>
> 9. Star Wars (SDI/BMDO) weapons systems have ironically been derived from extraterrestrial technology. They are "made in America" and elsewhere in the Group of Seven countries. These weapons systems include: killer lasers, neutral particle-beam weapons, kinetic- kill missiles, electromagnetic-pulse weapons, tactical nuclear weapons, and plasmoid pulsed laser-maser weapons.

In addition it was the opinion of the author in his conclusion of this research that...

> The ET's seek peace, human attention to reversing global pollution and the sharing of theological, historical, metaphysical, scientific, cultural and technological information. They also wish to be of assistance during upcoming and foreseen geophysical cataclysms and social turmoil. But only if we want their assistance.

In a further conversation with Dr. Wolf, the pressure applied by the Vatican concerning this forthcoming disclosure was revealed.

Official Within MJ-12 UFO-Secrecy Management Group Reveals Insider Secrets

http://www.jps.net/drboylan/wolfdoc.htm

Dr. Wolf told of learning during a high-level briefing in the UK that the Vatican had been applying pressure concerning the possible announcement by President Clinton of UFO visitation.

The ETs have other concerns, too. They discussed God and death with Wolf. And they pointed out that all worlds are connected. "One Hiroshima atomic bomb can affect millions of cultures in the galaxy." And they told him, "Thought is energy. It doesn't stop at a galactic 'barrier', and is received on other worlds." Because of what he learned, Wolf stated, "We are at a crossroads. The issue is to give a viable future to our children." He notes, "Humans are beginning to change, to evolve, and are looking for spiritual roots. There is more to God than people get in church."

Wolf is convinced that it was psychotronic directed-energy devices which were directed by rogue elements within the UFO Cover-Up organization at certain UFO researchers pushing very strongly to end the UFO Cover-Up. These include: the Director of Committee for the Study of Extraterrestrial Intelligence (CSETI), Steven Greer, MD, his principal assistant, Shari Adamiak, Congressman Steve Schiff, who demanded UFO documents from the Air Force, and USAF Col. Steve Wilson, who revealed his heading Project Pounce, the UFO retrieval unit. Already Colonel Wilson and Shari Adamiak have died of their cancers, and Congressman Schiff has had to terminate his political career. Now Dr. Wolf has received a medical report which suggests a possibility that he may have cancer. It is not yet clear whether he will still have access to a certain extraterrestrial honey-like substance, which he took several years ago for his metastasized colon cancer, and which brought that episode into remission.

Concerning the U.S. government's slow pace of disclosure about UFO reality, Wolf commented, "The government is doing a balancing act, because if Free [Zero-Point] Energy and all the ET technology come out all at once, it'd hurt the

> stockholders [in obsolescent industries]. The multinational corporations don't want to lose their power.

While an additional later in 1999 came to light on further disclosures from the UFO/ET's studies group.

http://www.jps.net/drboylan/wlflk1.htm

> The UN maintains an unacknowledged Special Intelligence Branch, which is involved in working with all countries and their intelligence services around extraterrestrial matters. Dr. Wolf revealed that he is negotiating with an official within the UN's Security Council staff to have the UN release UFO information it possesses in 1999. This official has advanced degrees, and has been given "incredible powers" during her ET encounters. He says, "The UN people know everything (about the visiting ETS). And (Apollo astronauts) Edgar Mitchell and Gordon Cooper have made appeals to the UN to let the truth be known."
>
> He added that a "Cabal" is a significant part of the problem. The Cabal is a rogue group of private industry, intelligence and military officials who have aggressively misused national security secrecy to achieve illegal monopoly over extraterrestrial technology. The Cabal is so entrenched, that it is hard for official government to remove them. Some of the Cabalistas are very high military officers, while others hold leadership positions as Senators and Congressmen, predominantly but not exclusively of Republican party membership. "They want Clinton out of the loop [of detailed UFO/ET information]," Wolf said, because the President has given private indication of being prepared to make public announcement of UFO reality as his departing message, if he is forced out of office. Wolf stated that "the Cabal is slowly dying off, and some are being relieved of duty." When I retorted, "But not fast enough," Wolf sighed and did not disagree.

In addition:

> Dr. Wolf became quite excited as he shared with me signals he has been receiving recently, that "Something is coming, so *big* that the government will *have to* announce UFO reality." When I pressed him for a timeline for the big event, he replied, "six months to 1-1/2 years."

And finally in June of 1999 it was also reported that this acclimatization programme of ET interaction had been brought forward with....

> An end-of June, 1999 telephone conversation with Dr. Michael Wolf, a consultant to the National Security Council's Special Studies Group, brings fresh news and encouragement about the government's Acclimation Program of preparing the public for official acknowledgement of extraterrestrial contact.
>
> Dr. Wolf has been in communication with two top officials of the Special Studies Group, (popularly known as MJ-12), a General and an Admiral. Wolf has been trying to convince General "Bob" and Admiral "X" that a total Cover-Up is no longer needed, and that the public is ready to handle news of ET reality.
>
> Now Dr. Wolf reports that his pushing has been successful, and has resulted in these top officers' authorizing a speed-up in the Acclimation Program. Dr. Wolf also excitedly shared that this effort "has support in very high places" in addition to the Special Studies Group.

And THE final say?

FINDINGS ABOUT THE REAL ETs
VERSUS THE BOGEYMEN OF PROPAGANDA

http://www.jps.net/drboylan/realets.htm

As to whether the ETs will "save the day" as regards Earth's ecological crises or the coming cataclysms know as Earth Changes, it is up to us, and the ETs tell us that, to save our own day (and planet). They will help, but only if asked, (and not shot at.) My analysis, and that of Drs. Mack, Sprinkle, Fiore, et al., is based on behavior (extraterrestrial) over time (centuries). "Actions speak louder than words."

Actions speak louder than words !

Of the two possible above scenarios, we choose to completely by-pass the one offered up in the letter we received, leaving us to ponder on the process that is being eluded to in the latter information. **IF** it is the process that has been chosen, to which we are attempting to show some **POSITIVE** signs of change to this **ET** monitoring group...People **it stinks!**

We have a military industrial complex calling the shots on this scenario. This makes absolutely no sense to anyone. Firstly, industry works off a profit/ loss sheet, and they have made a huge profit in providing the military with **Nuclear** and **Biological** weapons and equipment. Therefore, losses they do not want, yet it is a huge loss that they are faced with. No one wants to direct any of their profits towards research to find a way to neutralize these ageing stockpiles of **CONTAMINATED WASTE**.

Next, is the military machine, and like any other properly manufactured machine, you ensure it has all the required parts (in good working order) and then assign it to carry out it's programmed task.

First let us show some good will, show the other side that we have nothing to hide... let them have a very clear view (from their satellites) at our facilities. Cut down or remove a few trees, cut the grass etc., or may be we will get really lucky and have a freak fire that will clear the field of vision around 1 or 2 of our facilities. Then who knows we may get really lucky and an earthquake will occur right where we have all that old troublesome stockpile of **Nuclear and Biological waste.**

BOY WOULD WE BE LUCKY!

Otherwise people (or our **ET observers**) may think that we just could not clean up the mess in our own backyard...otherwise, if they thought that they might not let us play in theirs.

SURE BEATS DUMPING ACID INTO AN ALREADY DYING ECO-SYSTEM!

SURE HOPE THE OTHER GUYS UNDERSTAND HOW FORTUNATE THEY WERE TO HAVE THAT SAME BURDEN REMOVED, AND HOPE THEY DON'T THINK THAT WE HAD ANYTHING TO DO WITH IT AND HIT THE WRONG BUTTON...

A "STARRY" BIRTH EVENT

The more we learn about our environment, our origins in the cosmos and the evolution of the stars and planets around us, the more we realise how little we really know!

Until very recently, 1999, scientists assumed that the planets circling our sun as being the ones they were born in, however, it is now realised from research carried out on the planet of Jupiter that this may not be the case. I wonder if they read our articles we placed on the web some months prior to this!!

Their conclusion was the following article;

> **Jupiter gave birth to Uranus and Neptune**
>
> **By BBC News Online Science Editor Dr David Whitehouse**
>
> New suggestions that the planets in our Solar System have not always been in their current orbits have been put forward by two teams of astronomers.

This work, along with recent speculation that Jupiter may have formed much further from the Sun than its current position, and the discovery of other planetary systems orbiting other stars, is forcing a reappraisal of our understanding of how the planets were formed.

Writing in the journal Nature, researchers from Queens University in Kingston, Canada, propose that all of the giant planets in our Solar System formed in a narrow region of the gas and dust cloud that surrounded the early Sun. They suggest that they ended up in their present orbits as a result of violent and chaotic scattering.This would mean that when Jupiter, our Solar System's largest planet, formed, it triggered the birth of other giant planets nearby. In a way, Jupiter was the "midwife" of the Solar System.

Rocky cores

The four major planets in our Solar System are classified into two "gas giants" (Jupiter and Saturn), that have a small rocky core surrounded by a large hydrogen and helium atmosphere and also two "ice giants" (Uranus and Neptune) that have icy mantles around their cores and only a thin atmosphere.

> Scientists have always been slightly puzzled by the positions of Uranus and Neptune because in their present locations it would have taken longer than the age of the Solar System for them to form.The scientists from Queens University suggest that the four giant planets started out as rocky cores in the Jupiter-Saturn region, and that the cores of Uranus and Neptune were tossed out by Jupiter's and Saturn's gravity.
>
> In the simulations, the ejected planets went into highly chaotic orbits for a few hundred thousand years after which they settled down and gradually migrated to their present, nearly circular orbits.
>
> **Stable orbits** Another group of scientists, also writing in Nature, from the University of Toronto, have simulated how planets such as Jupiter may have formed in the first place.
>
> They found that gas and dust circling the early Sun that starts to accumulate to form a proto-Jupiter creates a spiral density pattern in

> the surrounding disk material. The proto-planet accretes mass rapidly through the spiral arms but when the planetary mass reaches four-to-five-times Jupiter's mass, the disk rapidly fragments into smaller proto-planets. Over hundreds of thousands, or millions of years the proto-Uranus and proto-Neptune would be flung outwards by the now smaller proto-Jupiter'sgravity.
>
> Not too long ago, scientists regarded the orbits that the planets circle our Sun as being the ones they were born in. Now they are realising that this is not the case.
>
> Uranus and Neptune may have migrated outwards and Jupiter may have come in from the outer cold.
>
> One of the questions scientists would like to answer is whether the Earth has always been where it is now?

The above then questions the idea that Jupiter may have formed within this solar system, thus forcing a reappraisal of our understanding of how the planets were actually formed. Therefore suggesting that the planets in our solar system may not have always been in their current orbits, begging the question, which we answered last year in our series of articles, whether the Earth has always been where it is now?

Well, perhaps they did not read our article after all, where in the Giza - Genesis connection series we offered the explanation of the replacement orbital pattern between Earth and Mars. That is, Earth was where Mars once was.

We are living in an evolving universe, matter is constantly on the move and evolution is a continual process, and because of this stars and planets are constantly being born and generated. However, for a clearer understanding one may also take into account the effects of time as to their present placement, which is covered in some detail throughout the book.

To see an example of this one only has to look to the number of recent photographs taken by the Hubble spacecraft such as **PILLARS OF**

CREATION IN A STAR-FORMING REGION (Gas Pillars in M16 - Eagle Nebula) where the structures shown are actually columns of cool interstellar hydrogen gas and dust that are also incubators for new stars.

With advances in Infra Red detection illustrated by the image, we now have a method whereby we can peer directly into the centre of such stellar birthplaces.

Our sun, which we have explained in our previous articles, is a product of a greater star in the Pleiades region in the constellation of Taurus. Every 26,000 years or so, procession, our star aligns with its point of origin, as it did so quite recently. In addition, our sun, like many other stars in the cosmos, has so much mass that it is able to produce its own light, this being the feature that distinguishes stars from planets. As many may have observed, over the recent months many anomalies have been witnessed relating to our sun including the continual and gigantic active CME's (Coronal Mass Ejections).

In volume 1 we wrote a chapter entitled 2012, COMET LEE AND THE BIRTH OF OUR NEW NEIGHBOUR_explaining the soon to be released body from our sun, the actual birth of a new planet and our new neighbour. Judging by what 'Science' now understands, which really is very little as we have seen, we think that we can justify that indeed a planet can be born, and that very soon we will witness this very scenario on our own back door.

Always remember that our SUN, like all other astral bodies has been Evolving since the time of its own conception. The closer we move towards the December 21st 2012 date the more we notice the changes which the Sun is undergoing as it prepares itself for an adjustment at that time.

Chapter 10

TIME AND TIDE WAIT FOR NO MAN

Throughout the year 2000 and the early part of 2001, a plethora of events occurred which have both amazed and mystified a great number of observers. Many of these events we have attempted to explain in relation to the coming of the climactic earth changes of 'mother earth' as she goes through this phase before the final 'birthing' event.

In addition, we have offered a consistent thread from the origins and purpose of the layout of the Giza Plateau Pyramid complex, right through to the various scenarios we now find ourselves witnessing. Scenarios that independently may at first be taken out of context, however, once placed together form a gigantic jigsaw of the most bizarre and frightening dire contrivance ever constructed this side of the Galaxy.

The connections once made have been obvious, and include such subjects as the anomalous solar activity, Brookhaven, HAARP, Chemtrails, earthquake activity and more.

In all our previous work we have offered our caution into the real agenda of such coming events as the great disclosure of extraterrestrial activity, the false prophet and what actually we should be looking out for. Naturally, a few have equated our advice for the coming months ahead as being part of a great mis or dis- information plan. We assure all of you out there nothing can be further from the truth

To disregard totally our research and the findings it brings with it would be to miss out on the greatest opportunity (ascension) in the history of this planet, one that may not come around again for at least another 500,000 years. We have this one great opportunity NOW, as they say "USE IT OR LOSE IT".

For those who do not accept what is occurring, well, sit back and watch in awe and wonder, for events WILL prove out.

One final point before we get down to the real nuts and bolts of these issues, a few people think that we offer this information solely to those living on Terra, planet earth. Would it be such a surprise to learn that the information is also intended for others! This is an opportunity for those off world too, remember us "earthlings" are relatively new kids on the galactic block.

HISTORY HAS A HABIT OF REPEATING ITSELF

On September the 13th, 2000 an extremely large inbound object arrived on a direct collision course with earth, this coincided with the arrival of part of the debris from the Mars/Phobos event that has been discussed in the media of late, as the following article demonstrates. (See also chapter 3)

Phobos Impact Scenario By James van der Worp
YOWUSA.COM, July 13, 2000

> On June 5, 2000 comet 76P/West-Kohoutek-Ikemura had its closest encounter with the planet Mars. This event has generated a wide-reaching debate on the Internet as many feel that this event was prophesied by Nostrodamus, Scallion, Gilbert and Mother Shipton.
>
> While NASA had originally planned to observe the flyby with the DSSI space craft, their position is that nothing happened and that the 76p flyby is a nonevent
>
> Following NASA's statement that "nothing happened" the interest in 76P began to fade. However, 76P has not been sighted since the June flyby of Mars and NASA is still publishing 76P data based on pre-flyby sightings. Since then, the Earth has been repeatedly buffeted by a outbreak of violent

solar storms. Consequently, NASA's inability to sight 76P plus the current solar activity has refocused attention on the prophecies of Nostrdadamus, with specific regard to C10:Q72.

As a member of the yowusa_space@egroups.com Special Interest Group, I've been watching all of this detached interest. I'm not one for jumping to conclusions or wild speculation, but what began to trouble me most was NASA's dismissive and condescending tone with regards to this matter, despite the fact that their data is all based on pre-flyby observations. To paraphrase Shakespeare, NASA say that this speculation is *Much Ado About Nothing*. However, I am beginning to think that NASA has begun to "Protesteth too much."

Given that NASA's odd behavior with regards to this matter had piqued my curiosity, I decided that I would see if I could construct a Phobos impact scenario, given that Phobos was thrown out of its unstable orbit around Mars sometime during the 76P flyby and is now headed towards Earth.

At first, this project was a interesting diversion no different than working a crossword puzzle, but as I began to run and rerun the numbers, a chilling possibility began to emerge and I realized that given an extraordinary sequence of coincidences, there is a remote chance that it could happen as the prophets have warned us.

Keep in mind that visual data is either unavailable or being withheld by the government, so all we can do is to present a mathematical assumption. Granted, numbers can lie, but it is important to note that the Planet Pluto was first discovered mathematically and then sighted some time later. In this case, the same process could apply to Phobos, a unstable moon of Mars, that has been thrown into the oncoming path of our own planet.

Does the timing of these events correlate to a cause and effect scenario, or can it be that within this universe there are in fact various "time" frames. In addition, this begs the question, if an action or event occurs in one time line, does it actually create a simultaneous event in another, and if so when?

Some of the debris regarding the above inbound events may be quite large, and the associated dust cloud and debris will also be coming on a direct collision course with Earth. However, people should NOT get caught up in the fear of these imminent impacts. It is believed that these inbound objects will not reach the EARTH, but some of the smaller debris may find their way through, for this cannot be avoided.

ALL of these inbound objects will be "bounced" off the grid, which was activated on the 5th May 2000 and was further strengthened on the 12th May 2000. This is the very same grid that has been defusing the massive CME's (Solar Coronal Mass Ejections) that have, and continue to slam into earth's ionosphere, only to poop-out before they reach the surface.

"HE" understands that there are those non-believers, although they are doing their best, they just do not have the insight to see all they should see. "HE" loves these people very much, and encourages them to question and to learn more. To and for these people he has provided a few visual aids, these aids were provided, as usual, through Crop circle formations.

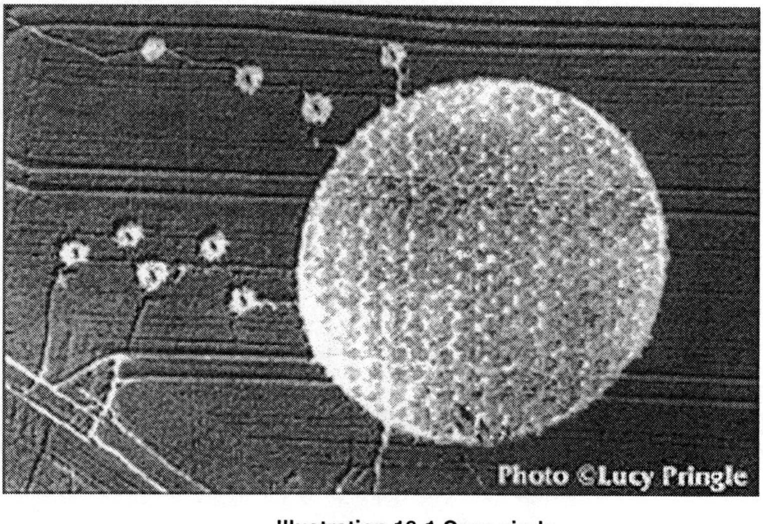

Illustration 10-1 Crop circle

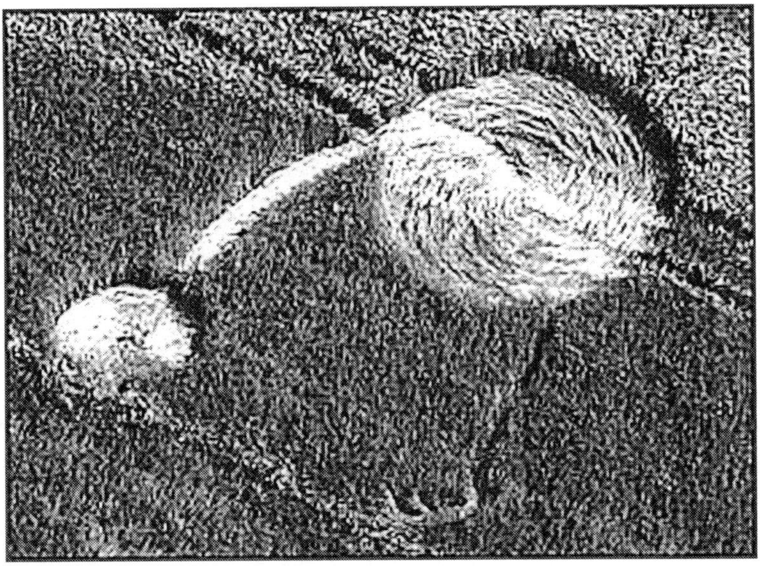

Illustration 10-2 Crop circle

In the above Crop Circle you will notice that the objects are on a collision course with the Earth's shield and grid. One might wish to pay attention to the direction, in addition to the approximate angle of these inbound objects.

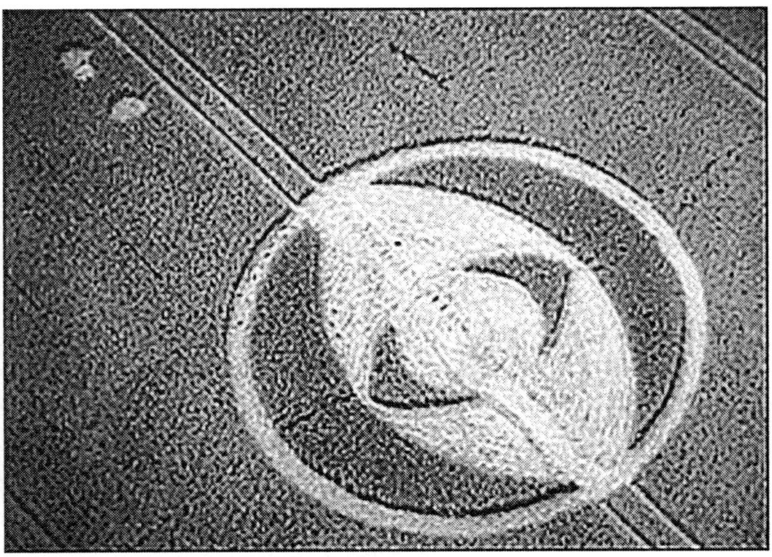

Illustration 10-3 Crop circle

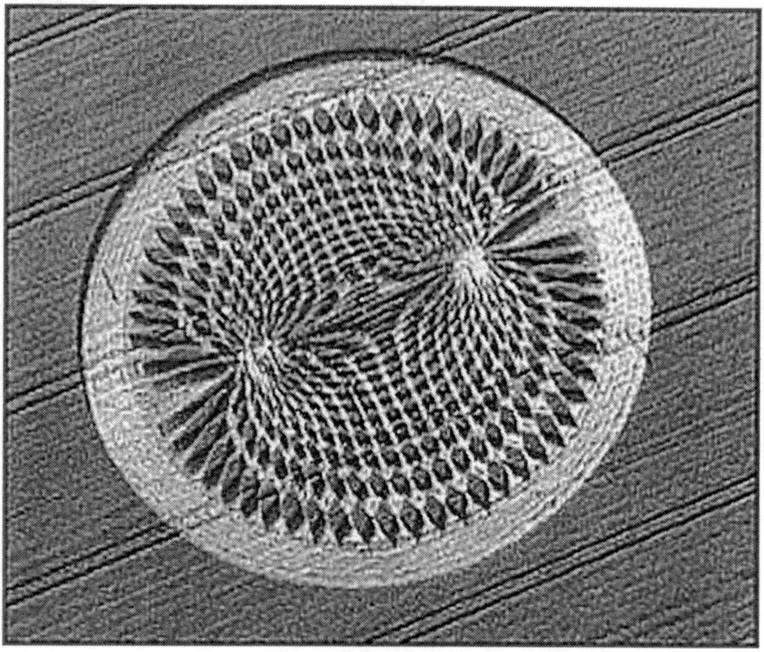

Illustration 10-4 Crop circle

In the Crop circle above you will notice that the shield of the grid itself remains intact. Now check the dates of the crop circles, the first crop circle (the one showing the inbound about to impact with the shield) was given before the second crop circle crop circle (the one which shows the shield intact but the grid shuddering or vibrating).

In previous work we spoke of future earthquakes, volcanic eruptions and tsunamis, etc. this imminent impact to the grid will create some or all of the above. There is no way in which these events can be prevented as far as we are aware. However, the grid itself will be absorbing some huge impacts and this in turn will create a need to re-disperse the energy created due to this impact, thus the shuddering, or vibrating effect shown in the second crop circle diagram.

For those who are aware of the physics involved in such an event, relax and do not be concerned that the impact will be throwing Earth off it's existing orbit, the situation as we understand will be remedied and under control.

Although we have chosen to use only but two examples of the recent crop circle formations in this particular section, there are many others and they may continue to emerge in various locations. Therefore, "HE" will keep showing those who cannot, or are reluctant, to see that information for which "HE" is trying to forewarn "HIS" people.

EPHRAIM

The Book of Revelation is not open for discussion or for man's interpretation, it is what it is, that being, the future history of events that will transpire. The point of no return was reached at the time the spear was jabbed into the Christ at the time of his Crucifixion. For the Christ and the Earth are but the same and we have continued to abuse this dying Earth. For although Christ sacrificed himself to try and save the Earth and his People, Christ, like the Earth is now dying, while the Rose (God's blessing) has now been passed over to Joseph's second born, EPHRIAM.

If we were to lay eyes on Ephraim at this point in time one will see only the body of a child, but his face reflects the awareness that he carries the knowledge and strength of conviction of a much older, wiser and mature learned person.

THIS MESSAGE FROM EPHRIAM

> Send my message to all those who now suffer. I am here to bring FREEDOM. FREEDOM and FREE WILL are one. A message also to those Tyrants, "you know who you are" you will be suppressed as you have suppressed others. Be warned...the time is now. The choice is yours. The choice is FREEDOM

FROM OSIRIS...

> We love you...regardless of everything else we will push through the Love

One of the inbound, that came in on or about the 13th September 2000, is believed to be none other than WORMWOOD! [1]If you recall, when we wrote the chapter on the Great Pyramid, THE ARK and the EXODUS connection, the term 'wood' in the book of Exodus was used when making reference to 'rock'. Now we have ships hiding in and around the inbound, hence the term burrowed, or 'wormed' inside. This scenario (of 'worming in') may be read in a posting of the stsci.edu public information site, embargoed until July 28th 2000: *Hubble Sees Comet Linear Blow its Top* at

EMBARGOED UNTIL: 1:00 a.m. (EDT) July 28, 2000

Hubble Sees Comet Linear Blow its Top

> Lackluster comet LINEAR (C/1999 S4) unexpectedly threw astronomers a curve. Using the Hubble telescope, researchers were surprised to catch the icy comet in a brief, violent outburst when it blew off a piece of its crust, like a cork popping off a champagne bottle. The eruption, the comet's equivalent of a volcanic explosion (though temperatures are far below freezing, at about minus 100 degrees Fahrenheit in the icy regions of the nucleus or core), spewed a great deal of dust into space. This mist of dust reflected sunlight, dramatically increasing the comet's brightness over several hours. Hubble's sharp vision recorded the entire event and even snapped a picture of the chunk of material jettisoned from the nucleus and floating away along the comet's tail.
>
> *Credits: NASA, H. Weaver and P. Feldman (Johns Hopkins University), M. A'Hearn (University of Maryland), C. Arpigny (Liege University), M. Combi (University of Michigan), M. Festou (Observatoire Midi-Pyrenees), and G.-P. Tozzi (Arcetri Observatory)*

It appears that Linear seems to be "misbehaving" and indeed on May 20th 2001 we observed the following news bulletin from Spaceflightnow.com;

> **Sunday:** May 20, 2001
>
> **Comet LINEAR splits further**

New images from the European Southern Observatory show that one of the two nuclei of Comet LINEAR, now about 100 million km from the Earth, has just split into at least two pieces. The three fragments are now moving through space in nearly parallel orbits while they slowly drift apart

This comet will pass through its perihelion (nearest point to the Sun) on May 25, 2001, at a distance of about 116 million kilometres. It has brightened considerably due to the splitting of its "dirty snowball" nucleus and can now be seen with the unaided eye by observers in the southern hemisphere as a faint object in the southern constellation of Lepus (The Hare).

Comet LINEAR was discovered on January 3, 2001, and designated by the International Astronomical Union (IAU) as C/2001 A2.

Six weeks ago, it was suddenly observed to brighten. Amateurs all over the world saw the comparatively faint comet reaching naked-eye magnitude and soon thereafter, observations with professional telescopes indicated the reason for this strange behaviour: the comet's "dirty snowball" nucleus had split into two pieces.

During the splitting of the nucleus, fresh material from the interior of this frozen body is suddenly exposed to the sunlight, causing a rapid increase in the evaporation process. More cometary material is released and the overall brightness increases, as more sunlight is reflected off the dust around the nucleus.

The VLT observes three fragments

But Comet LINEAR has just shown that it is good for another surprise. When astronomers at ESO's Paranal Observatory turned the 8.2-m VLT MELIPAL telescope (UT3) towards that object in the evening of May 14, they noted that one of the two pieces of the nucleus appeared somewhat elongated. The comet is rapidly approaching the Sun — it will pass through its perihelion (the point closest to the Sun) on May 25, and it was quite low in the sky (about 20 deg above the western horizon). Accordingly, the image quality was not perfect, but there was no doubt that something was going on with the fragment that was closest to the Sun (denoted "B").

And indeed, when the 8.2-m VLT YEPUN telescope (UT4) obtained another image of the comet in the evening of May 16, it was obvious that fragment "B" had split into two. In fact, the astronomers suspect that there may be other, smaller pieces.

Therefore, it is believed embedded and dug into this inbound are the forces of the Anti-Christ. They, like the inbound, will be "Bounced", (see chapter on the activation of the magnetic shift) however, the problem that arises with this scenario is that they will then be within striking distance of EARTH.

In addition, astronomers attempting to learn more about the demise of comet Linear were hampered at the most important viewing window of opportunity, by a forest fire that threatened the cluster of telescopes atop La Palma in the Canary Islands.

An interesting and relevant quote may be found in Luke.

> And he [Jesus] said unto them [the 70 disciples], I beheld Satan as lightning fall from heaven. (Luke 10:18.)

The Anti-Christ has called in his forces, and although he even walks amongst us here on Earth, he has yet to come forward. The false prophet has yet to clear the way for the Anti-Christ to step forward. This scenario is now set to unfold and make no mistake the Anti-Christ is very good at what he does. Because of this, the coming months will indeed be a very trying time, especially as the Anti-Christ attempts to influence your free will choice.

WE are unaware of his overall plan, and will not be aware of it until the Anti-Christ makes himself known, however, we do know the way the false prophet will begin to set the stage for the Anti-Christ.

As Revelation informs.

> And [after the war in heaven] the great dragon was cast out, that old serpent, called the Devil, and Satan, which deceiveth the whole world: he was cast out into the earth, and his angels were cast out with him." (Rev. 12:9)

FREQUENCY AND VIBRATIONAL STATES

The first point to understand is that everything that does exist does so by being held in its state on specific harmonic, or vibrational levels. ALL things at a fixed vibrational frequency will exist in a given time frame. Once all objects in the same time space are at the same vibrational frequency, they can be treated as one singular object, or, in our case, they can be controlled as a singular unit.

Therefore, there are those that would, and are, act in an attempt to take advantage of this situation. Hence the reason why most known alien races are telepathic, for once you become "tuned in" you then arrive at the point where your thoughts and freedom are no longer one's own.

Once this process is initialised you are then at the mercy of the one's who control that frequency. It is through, and as a result of this that they may control and alter one's beliefs and history and mould you into what they would have you be, **and the first point they would tell you is that there is no God.** For should one even suspect that there was a GOD, they would have to run the risk of the discovery that the freedom of personal growth would attract some of their followers (or slaves) away from their control.

It was this very same control possibility that the New World Order had in mind when they discovered the Zeta's possessed this knowledge for them to carry out such a plan.

The Zeta's found us humans a very interesting race of people, for they had never encountered a race that was driven by emotions to the extent that we are. Therefore, the deal was made and the chemical cocktail drawn up. It was then left to the NWO to begin administering this 'control' system en-masse to the people. This chemical cocktail has been designed to lower our vibrational rate, carrying with it some concerning side effects, which we will address separately.

This false prophet we speak of, at the time of writing, is "in ASHRAM" [1]nd when ready, will have reached the level at which he will have mastered this very same vibrational frequency. He has previously stated that when he appears on camera he will not speak, but that you will know his thoughts.

The only piece of advice that we can offer at this point in time is **do not look directly into this man's eyes.**

As far as breaking free of this much lower vibrational frequency, we suggest that you learn to generate a self-induced vibrational change (this was the one thing that the Zeta's could not comprehend - hence the fascination with our species). This is possible if you can induce an alteration to your emotional state, as such changes cause your vibration to alter accordingly. Any such change of emotion will alter your vibrational frequency, therefore, should you begin to feel somewhat lethargic or absent minded etc. For example one should create an emotional feeling that you feel very strongly about, such as anger, which is an easy emotion to bring to a rapid boiling point, especially when all you have to do is think of what they are trying to do to the people on this planet. Compassion and LOVE will also create a rapid alteration in your personal vibrational space.

In all the teachings man has learned throughout time, we have always been led to believe that there are only good, or bad, right or wrong etc. Unfortunately, there is much more to this than meets the eye, for there are varying degrees of right or wrong, good or bad. ONLY GOD has risen above all of these levels of understanding the real truth and "HE" cannot come down to us, WE therefore, must raise ourselves up towards "HIM". All "HE" can do is show us the way and ensure that all will be done to help those who wish to follow "HIM", as long as one lives to the Covenant which we have with "HIM".

As of RIGHT NOW, we are entering into a very trying and confusing time indeed and at this point in time the war is working its way towards the active part of its climax.

Cautiously, the various groups have begun to instigate deployment of their strategies and already the counter strategies are making themselves evident.

The **ONLY** winners here will be, after all is said and done, those who make ascension

FREE WILL CHOICE IS AN OPTION THAT ALL POSSESS THE RIGHT TO EXERCISE, AND THE MORE ONE LEARNS AND UNDERSTANDS, THE BETTER HE OR SHE IS OF MAKING A TRULY FREELY DECIDED CHOICE.

In this dimension, it will appear that the "good guys" are the Galactic Federation, and this is true, at least when you compare them to the alternative. For all those who will make their choice and do not choose to follow the one and only God, we hope that you at least choose the side of the Federation. Since the beginning of time the numbers and struggles of the Federation have increased, remember that they too (when they reached this point in their natural evolutionary process) had to make their free will choice. Their souls have remained here under the control of "The Mind Controllers", those being the SIX members of the upper Galactic Federation Council. If you will recall, in a previous section, we wrote on the fact that an extremely lethal potential situation had arisen, one that would have involved the use of nuclear weapons. At that time, due to the severity of the situation these six mind controllers met on another planet of our Solar System, this being not a social gathering as one could well imagine.

Bear in mind, presently there are approximately seven billion souls on this planet and each and every one of those mind controllers wish to get their share of the "Harvest". The more souls they are able entice to follow them, the greater the strength of their domain becomes, ultimately extending outward to other sectors of the Galaxy.

In periods of recent times, various Earth leaders have been approached and on each occasion received offer of help from these various groups. This is one way in which the Zeta's introduced themselves, always with their selfish motive, "they need people".

Here on Earth it was relatively straightforward for these groups to infect our societies, because here man's nature has been gearing up to be one of power over others. This was the catalyst that sparked the run for the NEW

WORLD ORDER. These people began to follow and worship that which they hold higher than GOD himself, that being the false god of power and greed. These very same people are now caught in a real catch 22 situation that being, the fact they themselves in turn have been duped.

To make matters even more confusing for themselves, they find that there **is** a GOD and “HE” has sent “HIS” people here to prepare and ready “HIS” people for the trials and tribulations that will shortly be upon us.

The main reason why the grid was placed in position, and why the polar magnetic shift will come about will be looked at in detail in a later chapter. However, in addition it may be referred to as the fulfillment of “his” part of the Covenant.

In a few months this grid will be ripped apart by the very forces which gave it its strength, and at that time the real battle will begin in earnest.

“HE” also established the grid to show others and to remind them that there truly is only one God, and because he is an all-loving God “HE” will not forsake any of his children, including those who may have in the past made mistakes and temporarily strayed from “HIM”.

It is the test of the weight of the feather (see our chapter “A Given Sign”) that will determine whether those who passed by the opportunity previously, will make ascension this time around, that being the END TIME.

REMEMBER THAT GOD CANNOT COME DOWN TO US, HE CAN AND WILL HELP THOSE WHO SO REQUEST, PULL THEM UP TOWARDS HIM. ALL HE ASKS IN RETURN IS THAT YOU HOLD TO YOUR FAITH AND LIVE WITHIN THE PARAMETERS OF THE COVENANT THAT HE HAS WITH US.

IN DOING THIS HE TEACHES THAT WE SHOULD GIVE OUR LOVE FREELY TO THOSE WHO WOULD TRY TO HOLD US BACK. FOR WHEN THE NEXT END TIME ARRIVES, IT MAY BE THAT MEMORY OF THE LOVE FREELY GIVEN, WILL HELP THEM IN THAT TIME WHEN ONCE AGAIN THEY WILL BE GIVEN THE OPPORTUNITY TO ASCEND.

THE CHEMICAL COCKTAIL

THE "CHEMTRAIL" CONNECTION:

One of the reasons for the effort to lower our vibrational rate is to cause lethargy and absent-mindedness on A mass scale, thereby creating thicker "veils" for us to try and see through. If we cannot see what is happening, then it stands to reason we cannot take action.

For many years our environment has continued to be polluted, our air, our water and our food in effort to lower our vibrational rate. As the time gets closer for the False Prophet to make himself known, additional methods are being employed.

We observe in our skies, jet aero planes constructing fancy designs ultimately spreading out to become cloud-like formations. Dubbed "chemtrails", these formations consist of long-chain polymers in which can be embedded other organic or inorganic compounds, such as viruses and bacteria. On days when the atmospheric conditions are less than ideal, the mixture does not break into small particles as it falls from the planes, but falls as if spider webs are falling from the sky.

These chemtrails work on several levels. The first, and most direct, level is a lowering of the immune system by constant bombardment of the body by bacteria and viruses in the mix. To be taken into consideration in this direct attack, is the fact that some people are allergic to various compounds in the mix. The human body, already immune-lowered due to pollutants in our environment, is unable to cope with this extra bombardment and will eventually succumb to illness and perhaps even death. A lowered immune system is a lowered vibrational rate.

The second level on which the chemtrails work is through the brain. Worldwide, people are reporting memory loss as a symptom of the illness following chemtrail activity. Short-term memory is a function of the beta portion of the brain, and it could be that "beta-blockers", usually used for depression illnesses, are included in the mix.

In the drug world, the beta blockers (examples: atenolol, metoprolol, propranolol) act as competitive antagonists at the adrenergic beta receptors.

The newer agents tend to be more selective for the cardiac (beta-1) receptors which allows for decreased systemic side effects. Most of the generic names for beta blockers end with "olol".

A further veil therefore, is added to our already dimming sight.

A third, and more underhand attack, comes from the inclusion in the mix of a "DNA splicer", normally used in laboratories. Genetically engineered food is becoming more prevalent on our grocery store shelves, and the long-term effects of the DNA changes made in the food chain on the human body are as yet unclear. It has recently been found that the genetic changes do cross species, such as from crops to honey bees, and to the intestines of rats and mice.

Consider that we now have in our system a DNA splicer, which could cause the genetic changes from our food to alter our own DNA. In doing so, our vibrational rate could be easily lowered through a DNA alteration.

The fourth, even more obtuse attack is the inclusion in this cocktail of aluminium particles, which may be electrically charged. Consider the effect of the human body infused with these particles constantly in an environment of electrical waves, electromagnetic fields, microwaves and x-rays (especially with our sun being in high solar cycle). This will eventually lead to illness. Now consider the effect when a pulse is sent from such facilities as HAARP or Brookhaven Heavy Ion Collider, the end result could be mind-control on a mass scale.

This aluminium allows HAARP to change or allow grid vibrations, which not only affect weather patterns, but an attempt to alter the earth's grid, which they believe will help them control and override "natural laws", or in effect alter natural time and space.

In other words, one can liken them to children with new toys, having no sense of responsibility or fore thought as to the potential damage and affect that could and will have on other astral bodies, including the peoples who inhabit them.

So, as we have seen the chemical cocktail attacks humans on many fronts, affecting body and mind, in order to affect our "sight" and to make our discernment less than ideal. The False Prophet, therefore, will have a rather easier task in "leading the masses" once he makes himself known. His way has been paved.

In addition, there is a possibility that the Pope may die prior to the upcoming polar magnetic shift, but in any event this will occur in the very near future. **Do NOT** be alarmed if this appears to be an unnatural event, as in reality will in fact be a natural event as his time has come to pass. Prophecy has revealed that the Pope to succeed Paul John II will in act be the Anti-Christ, or his representative.

The new Pope will bring the downfall of the church and will leave Rome. This is the Third prophecy of **FATIMA**, that of the Pope leaving Rome in ruins, where he will head to the mountains where many will die and the church ultimately will be in ruins. His blood and theirs will indeed wash the Earth, as it is the "offering" of blood.'

These events will occur as stated. They could occur prior to the event of magnetic shift, but we do not know the exact time line. However, should any events occur before that time, you will at least be prepared to understand this particular event.

(Endnotes)

[1] Wormwood, as we know it, originates from the bitter plant and the word has long been associated and employed figuratively for something bitter or unpleasant to experience, reflecting its best known characteristic, its taste. The Latin name for Wormwood is *absinthium*, and is known for its mood altering effects.

> And the name of the star is called Wormwood: and the third part of the waters became wormwood; and many men died of the waters, because they were made bitter.
>
> And the fourth angel sounded, and the third part of the sun was smitten, and the third part of the moon, and the third part of the stars; so as the third part of them was darkened, and the day shone not for a third part of it, and the night likewise.
>
> And I beheld, and heard an angel flying through the midst of heaven, saying with a loud voice, Woe, woe, woe, to the inhabiters of the earth by reason of the other voices of the trumpet of the three angels, which are yet to sound

Revelation Chapter 8 Verses 11-13

1. A secluded swelling of a Hindu sage and the group of disciples instructed there. 2. A religious retreat.

Chapter 11
COSMIC EVENTS THEY BEGIN

Sydney Australia – Location 33:55'00S 151:17:00E

On the 13th September 2000 at 05:30 (local time Sydney – or 12th September 18:30 Zulu – UTC) until 17.30 (local) the following stellar alignments occurred.

Constellation of Delphinus was at 270 degrees West and at -25 degrees declination.

The moon was at 265 degrees West and at +15 degrees declination. 25 + 15 = 40.

Add 40 to 15 degrees (the Moon) = 65 degrees declination.

Opposite at this point is Phoenix – at 65 degrees declination, at 265 degrees and 270 degrees West, minus 180 = 85 and 90 degrees directional opposite. So the sun was at 97 degrees East.

In addition to the above, the moon was about to enter the constellation of Aquarius at 05:30 (local Sydney).

Thus, at precisely 17:30 (Local Sydney) the charts showed the following.

The constellation of Delphinus will be at 65 degrees ENE, while the Moon was at its FULL phase, 105 degrees ESE, and opposite was Phoenix at 147 to 150 degrees SSE horizon.

The sun was at 285 degrees West. 285 – 180 = 105 degrees that placed the moon in Aquarius at 17:30.

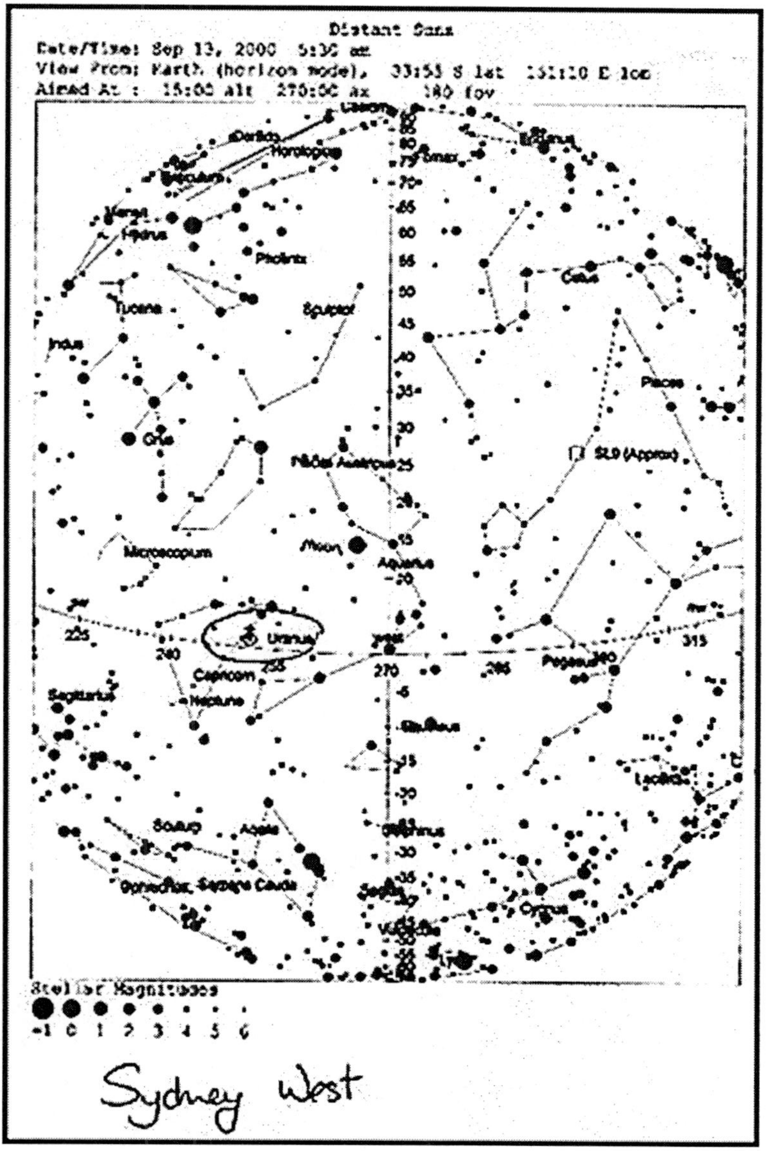

Illustration 11-1 Sydney sky chart September 13th 2000 AM

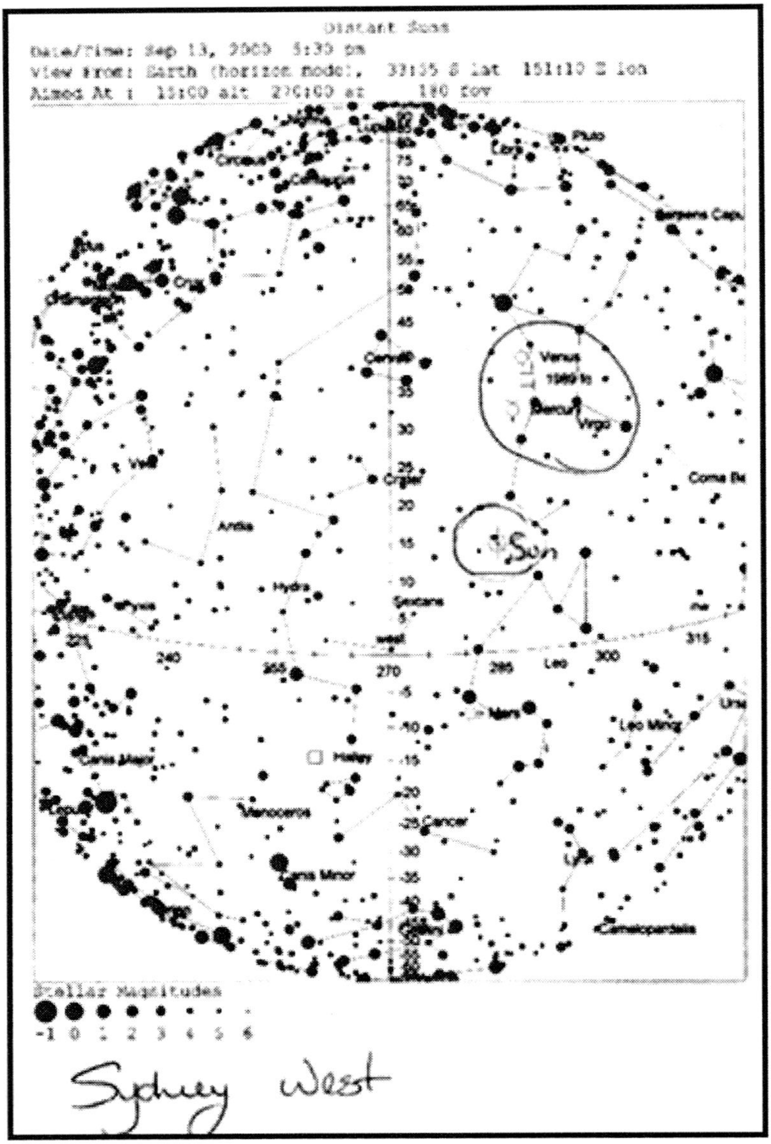

Illustration 11-2 Sydney sky chart September 13th 2000 PM

What does this mean?

It appears that the above scenario confirms the beginning of the cosmic events as laid out in one of our previous chapters, that is, the incoming of a comet, which will bounce off the grid, held in place around our planet. At least until 21st December 2000, realized in our time reference, 21st December 2001. (See chapter on Time)

The Moon.

The grid on the Moon is going to be strengthened and pushed outward, with provision for intersecting with the Earth grid. As the alignment with Delphinus (as one of the points of alignment) triangulates with one of the other points, Phoenix, these are subsequently re-routed around the Earth grid, or sent straight through to the sun. This then, creates the alteration to our space by altering the energy either way.

This effect, once the natural flow is interrupted then gives the sun a chance to cause a gap in normal function, subsequently allowing the window of opportunity to begin the internal reaction, and three months later the birth (of our new planet) occurs.

The above will be realized by the location of the monoliths placed on the Moon aeons ago, which will deflect the energy. These monoliths we dealt with in detail in Volume 1.

At the present time, these events will not weaken the Earth grid in anyway and no breach will occur.

To illustrate the above the following Crop Circles were observed during August 2000.

Illustration 11-3 Crop circle KILDEBROENDE – (Denmark) August 22nd

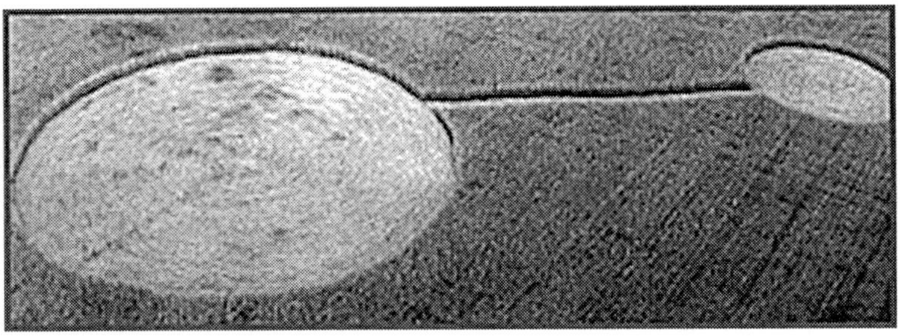

Illustration 11-4 Crop circle MOOSOMIN – Saskatchewan – August 30th

While two others were reported in MONTANA recently. The one on August 31st showing the change in energy flow –

Illustration 11-5 Crop circle Montana 1

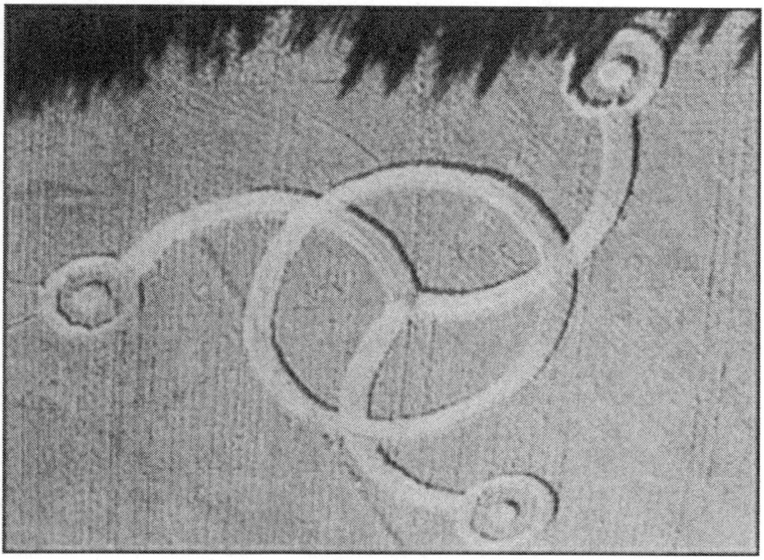

Illustration 11-6 Crop circle Montana 2

It is through this alignment that a series of events will soon begin to occur within our space. Communications will MOST likely be disrupted or cease completely, as a result of the changes that will take place once the energy within our space is altered. This being the result of the grids interacting once the effects of the magnetic shift comes about.

This is an extremely important time and many events will be realized, where there will be MANY disruptions to many events and meetings that we feel will be taking place at that time.

As people sit and observe these events that will occur, bear in mind that these events are necessary and being carried out to facilitate a much greater future.

Use this time wisely. Carry out some personal reflection and place your faith in HIM, for the weight of the feather is set upon the scales of Maat (truth and justice) and now is the time for the weighing of the heart.

Chapter 12

SPEECH OF THOTH

We have included this section with information that will hopefully answer many of the questions that have arisen to this point in time, and to prepare those for the coming months ahead. We trust that this information will give an air of hope and confidence in one's belief.

We referred in one of our previous sections to the anomalous signals that were being picked up and recorded by such groups as ELFRAD We believe, in addition from advice imparted to us, that the frequency of these signals are speeding up in order that a more efficient and quicker 'lift' (ascension) be expedited. This higher frequency is also more efficient as it utilizes less energy. In addition, the ships involved interact with the grid around the planet, which in turn interacts with all else, thus the double signal effect seen on the graphs at the time.

Our research group were informed that very shortly global evidence would be made available for all to witness, whether of the covenant or not, to demonstrate HIS ability.

This evidence materialized sooner than we thought and in a quite unsuspecting manner. We were originally informed to keep an eye on the satellite images, and naturally due to the interest in the anomalous solar images we assumed it would emanate from that quarter.

Well, imagine our surprise when we noticed the GOES - 8 IR Globe earth satellite image, dated July 10th at 17.45Z (UTC) while only two days later after being informed of the forthcoming evidence

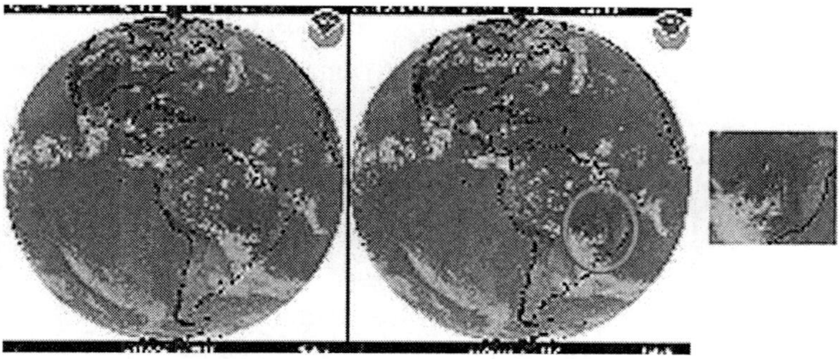

Illustration 12-1 of Satellite Image

It is interesting to note that we had been instructed to look for the image only days before, and strangely enough all the regular satellite sites at the time were in fact down, apart from an obscure university site that NASA presumably overlooked. Thus we were able to locate that one satellite image however, and it is our belief that "they" were expecting the image to be revealed would be that showing one of the Pick-up ships (used to pick up those in the book, and deliver them to the gate). Thus being the reason that this particular satellite image passed undetected.

We were told to place this image up on the web as soon as it appeared as a 'sign' so that all will know the **TRUTH** and be given **HOPE.**

It appears that what is known as the 'Rapture' may be indeed commencing. In other words, utilizing technology such as the ships, to take those whom are sent for and to a central location as it were. At this point those taken are received and their names checked against those in **THE BOOK**. Remember we talked of this book of records in previous chapters, some may refer to it as the **BOOK OF THOTH**, where records of all souls are kept.

If the person has lived by the covenant, they are allowed to enter, thus placing themselves in **His** care.

It was emphasized to us that the keeping of the covenant, a belief in the one, was of the utmost, and that any other god presented was not **Him**.

We will first explain, as we understand the situation, who and what group will be lifted initially, then go further, using the example of the ancient Egyptian "weighing the heart" ceremony, to demonstrate the how and why.

Those that will be leaving at this time will initially be the old and the frail, closely followed by the sick and the young children. Later, those not here to help with this specific time event, however there are many who will remain to assist and aid others who both are leaving and remaining.

Next, if there is time, those who have kept the covenant and have chosen to go either previously, or now. However, **it does not include those who came to carry out work at this time.** In addition, those who are physically deceased but still on this planet will also be taken as long as they are of the covenant and whose heart equals in weight to the feather

We will now explain the ceremony of the "weighing of the heart", towards an understanding how this scenario fits in at this time.

THE COVENANT
THE BALANCING OF THE HEART AND SOUL

Many have been enquiring as to the purpose of and the "qualification" for being of the **COVENANT**. In this section we will attempt to offer a brief overview in order that one may at least grasp some modicum of understanding, bearing in mind that this is an extremely complex scenario, where we do not profess to have all the answers.

Over the aeons of time our true ancestral 'origins', due in part to the many genetic interventions and variations, may never be fully realized. However, by reading through the ancient Egyptian and Biblical texts, and applying the analysis that we consistently followed, we may be able to open a crevice towards an understanding of this subject.

Before we submit our understanding of what people term the covenant, a few pointers may be helpful. We must initially point out that we do not really know who's names are written in the book of the Covenant, it is basically a personal thought on belief or faith. However, it is believed that in the main it would be the original tribes of what we call 'human kind', those possessing human DNA. In addition, those people who agreed to keep the spiritual connection, and have continually done so, including those who have asked to be saved and forgiven.

Which brings us onto our last point, those whose hearts balance out to the feather of truth, Mayet.

By examining the Biblical texts and comparing those of the ancient Egyptian texts for example, we may be able to see a substantial amount of similarities. Indeed, are they but the same record, but over the long period interpretations and translations confused and lost in the swirling mists of time.

According to ancient Egyptian "mythology" the beginning of our world rose from the primordial waste of waters, Chaos or Nun, known as infinity or 'nothingness'. It was said that Atum rose from Nun and known as the 'Father of the Gods' and often represented as standing waist deep in water holding aloft his hands supporting a solar barque (boat).

Continuing with our example, we have Nut, the sky goddess who was married to her brother Geb, the earth god, and as a result of their union, conceived two daughters, Isis and Nephthys, whom we shall now explain in this scenario of the covenant.

Nephthys was married to Set (Seth), but her loyalties were to her brother Osiris, god of the underworld and resurrection. It is understood that Nephthys bore no children with Set, so disguised herself as Isis and committed adultery with Osiris, consequently she bore **ANUBIS**, god of embalming. Nephthys abandoned Seth after he had brought about the death of **OSIRIS** (who split Osiris's body into many parts) lamenting over the death with her sister Isis. Indeed it was the custom at Egyptian funerals for two women to impersonate Nephthys and Isis and lament over the body of the deceased.

The baby **ANUBIS** was discovered by Isis in reeds in the Delta marshes of Lower Egypt and adopted him. A familiar theme one may ask?

The parts of the body of Osiris were scattered over Egypt, but found and put together by Isis and thus with the help of Anubis resurrected, while Horus, their son, took revenge on his father's demise and killed Seth.

This is the essential background to understand the part that Anubis, Osiris and Thoth play in the ceremony of the **WEIGHING OF THE HEART**, or **SOUL** upon death.

In ancient Egypt people who died were known as **OSIRIS X**, and once pronounced dead were judged in the hall of the judgement of the underworld, whereby the dead person was answerable to himself. Located in the centre of this hall, (the hall of two truths or double justice) would be a large weighing balance. Beside this balance would be Mayet, the personification of spirit of justice or of world order, and Anubis to ensure the scales are balanced and that the heart of the deceased used no trickery in the weighing procedure.

Thoth would be in attendance to record the outcome and to ensure the heart had a fair chance to justify itself, Anubis would weigh the heart against the ostrich feather of truth, Mayet, while below waits **AMMUT** (a hybrid, part lion, part hippopotamus and part crocodile) to devour the heart if not justified!

SPEECH OF THE DWELLER IN THE EMBALMMENT CHAMBER (ANUBIS):

Pay good heed, O righteous Judge to the Balance to support [the testimony] thereof. Variant: Pay good heed to the weighing in the Balance of the heart of the Osiris, the singing-woman of Amen, Anhai, whose word is truth, and place thou her heart in the seat of truth in the presence of the Great God.

The heart was considered the seat of intelligence and therefore responsible for man's actions and conscience. If the heart was equal in weight to the feather, truth, the soul was then proved innocent, thus the great Enniad of gods ratifying a favorable verdict. Thoth would then record the results kept in **THE BOOK**.

THE SPEECH OF THOTH:

From the ***BOOK OF THE DEAD*...or *PAPYRI OF ANI***

Thoth, the judge of right and truth of the Great Company of the Gods who are in the presence of Osiris, saith: Hear ye this judgment. The heart of Osiris hath in very truth been weighed, and his Heart-soul hath borne

testimony on his behalf; his heart hath been found right by the trial in the Great Balance. There hath not been found any wickedness in him; he hath not wasted the offerings which have been made in the temples; he hath not committed any evil act; and he hath not set his mouth in motion with words of evil whilst he was upon earth.

THE SPEECH OF HORUS TO OSIRIS IN INTRODUCING

ANI TO HIM:

Horus, the son of Isis, saith: I have come to thee, O Un-Nefer, and I have brought unto thee the Osiris Ani. His heart is righteous, and it hath come forth from the Balance; it hath not sinned against any god or any goddess. Thoth hath weighed it according to the decree pronounced unto him by the Company of the Gods, and it is most true and righteous. Grant thou that cakes and ale may be given unto him, and let him appear in the presence of the god Osiris, and let him be like into the Followers of Horus for ever and ever.

The meaning of the weighing of the feather:

Many will recall the Testament advice that there shall be no other gods before me or craven images and so on. This is because many have forgotten the Covenant with **Him**, they still have yet to believe in the one God**!!** Many follow or worship false God's and Deities, and yes, even Money is worshipped. Thus, this worshipping of these false God's adds weight to the heart...Until one truly lets go of these false God's their individual Hearts will weigh more than that of the feather.

Once these initial events have taken place, the **"SEVENTH SEAL"** will be opened and all those departing this time event will leave. Others will then arrive and the seal closed. You will know when this time has arrived as the signal that is being recorded by the **ELFRAD** group will have ceased.

It is believed that the sixth seal opening was the arrival of the comet Kahoutek [1]

(Endnotes)

[1] NASA Responds to Our Open Letter Regarding 76P/West-Kohoutek-Ikemura June 23, 2000

TO: Charles S. Morris of NASA
csm@encke.jpl.nasa.gov

FR: Marshall Masters
mail@millenngroup.com

Dear Mr. Morris

Thank you for the reply in which you raise a very significant point: "There has been no ephemeris posted on this comet because it is not currently observable."

While you did provide an updated ephemeris, is obviously based on a sightings of 76P made prior to the recent 76P flyby of Mars. The numbers therefore are based on assumptions as opposed to observed facts which is understandable. As you point out in your reply, "Hubble can not observe it either because it is too close to the Sun."

Once Mars, it's moons and 76P can be observed, then and only then can anyone claim with any certainty that this "much to do about nothing." As to the other issues raised in your reply, the readers can judge for themselves.

Marshall Masters
The Millennium Group

Chapter 13
PREPARATION

In chapter 11 where we dealt with the **September 13th Events**, we went into great detail as to the purpose and reasoning behind the heavenly alignment and cosmic events that were to unfold. We explained that an inbound would direct itself at earth (asteroid/comet etc.) but would "bounce" off the grid still in place around the planet.

This in fact did occur, however, many may have doubted or ignored the fact that this scenario did actually take place. Because it was not a direct hit, being bounced off somewhere in the region of the outer limits of the orbital path of our moon, the effects (as we described in that article) would be felt over a few days after the event, thus continuing to escalate for some time.

The reason why we mention this here is to demonstrate, utilizing as much as the available data as possible that this scenario did take place and the post effects culminating from the event that we are still witnessing today.

Because the major incoming was averted, or rather postponed, this has offered a limited time in which for certain people to prepare for what will be an interesting and eye opening period leading up to December 21st this year. A period of unsettling earth changes including extreme solar activity, earthquakes and a time when the Anti Christ will certainly make his move amongst the increasing confusion on this planet.

Immediately post September 13th a number of interesting cosmic events began, in fact so many that it would take a complete dissertation to list and explain all of the events. However, as a guide we list the main events, with the site references, in relation to our **September 13th** events (Cosmic events chapter) to demonstrate what actually occurred, and in fact is ongoing as one reads this article.

On September 12th, a full halo CME was observed and was thought to impact earth by September 14th – 15th. This was later termed the **Madonna CME** and by September 17th massive readings were observed on all the solar charts such as here on this article ***Solar storm smacks earth*** dated September 13th 2000.

> Scientists monitoring the solar blast have detected changes in its magnetic field that could cause trouble over the next few days.
>
> 9/15 12:15 PST IS THE EVENT ABOUT TO BEGIN?: comment added at 06:05 UTC on September 15: A solar wind shock was observed at ACE at 04:05 UTC. The interplanetary magnetic field has since then been northwards and only a modest increase in geomagnetic activity has occurred so far. Should the IMF swing southwards active to major storm conditions may occur.
>
> EDITOR: 9/16 Looks like the plasma went north, thus weakening the effect of the storm. Shields up?
>
> From Kent Steadman editor Cyberspaceorbit.com

While on the same day Charlie from the **Elfrad Group**, posted this note:

> **Would like to call your attention to a very unusual and strange event. The .9 hertz signal Elfrad has been monitoring for the past while has been joined by other transmissions. The regularly scheduled broadcast between 900 and 1000 hrs UT has some companions. The first signal burst started at 822 UT for a duration of 30 minutes. Then the regular daily signal began at 940UT for a duration of 57 minutes. At 1208 UT another transmission began with a larger amount of power. All three events were of the same frequency however each one had different amplitudes. Am unable to determine if originating from same location. Will**

update you in the event of further anomalies. Charlie, Elfrad Group

In addition from From: Hblondel at tampabay.rr.com (New Millennium)

The magnetic storm watch for tomorrow, September 15th, has been upgraded to the maximum storm watch values issued by the SEC. Previously the watch was for magnetic-A indices of >30 ... The replacement watch is for indices of >50, which should they occur will represent a major to severe magnetic storm event.

This watch is in response to two impressive CMEs that were observed on the 12th that appeared to be highly geoeffective [directed at Earth.] The latest advisory from the SEC suggests that the storm will begin very early tomorrow [UTC] which actually begins this evening in the USA.

Interplanetary shock detected passing ACE at 04:02 UTC on 15 September.

Solar Shockwave Heads for Earth

By Larry O'Hanlon,
Discovery.com News
Sept. 13, 2000 — The sun has thrown a powerful surprise shockwave of white-hot particles and radiation toward Earth, despite there being remarkably few sunspots on the sun.

The rare event, called a "great spotless flare," occurred on Tuesday and was apparently aimed straight at the planet. Its blast is forecast to cause a level G3 geomagnetic storm on the National Oceanic and Atmospheric Administration's Space Weather Scale. The scale tops out at G5.

The G3 storm will probably increase auroral displays in mid to high latitudes on Thursday night and Friday morning, but is probably not powerful enough to cause any major problems, said Ron Zwickl, deputy director of NOAA's Space Environment Center in Boulder, Colo. Super solar storms aimed just right can fry satellite components, disrupt radio communications and even cause power blackouts.

"A spotless flare is rare, but it does occur from time to time," said Zwickl.

Although spotless flares throw a kink in the process of trying to forecast space weather by monitoring sunspots — which are normally associated with such outbursts — they can be anticipated by watching for long stringy filaments of hydrogen on the face of the sun, says solar physicist David Speich, also of the Space Environment Center.

"A filament is a clump of solar mass — hydrogen — that sits on the sun," said Speich.

Sometimes those filaments break free of the magnetic fields that hold them in place, sending all that mass into space. That's what happened Tuesday when a filament about fifty times as long as Earth's diameter was seen across the face of the sun, then disappeared.

"This filament disruption is a fairly rare event," said Speich. "It's a tough forecast."

At the same time, solar observing satellites picked up a burst of energy from the sun — including a "full halo" coronal mass ejection that usually indicates that the flare was aimed right at Earth. It's similar to looking straight down the barrel of a gun and seeing the puff of smoke before the bullet hits.

Just how severe the geomagnetic storm will be when it hits Earth depends on whether the magnetic field of the shockwave is aligned with or opposite of the Earth's magnetic field. As anyone knows who has tried to push two bar magnets together, says Zwickl, there is a lot of trouble when like poles are pushed together.

While on the 14th September 2000

Quote...

> **Date: 9/14/00 4:07:03 PM Pacific Daylight Time**
>
> **From: Blondel – Tampa Bay (New Millennium)**
>
> **The magnetic storm watch for tomorrow, September 15th, has been upgraded to the maximum storm watch values issued by the SEC. Previously the watch was for magnetic-A indices of >30 ... The replacement watch is for indices of >50, which should they occur will represent a major to severe magnetic storm event.**
>
> **This watch is in response to two impressive CMEs that were observed on the 12th that appeared to be highly geoeffective [directed at Earth.] The latest advisory from the SEC suggests that the storm will begin very early tomorrow [UTC] which actually begins this evening in the USA.**

On the 16th it appeared that the plasma storm veered north, did the grid divert this incoming solar storm?

It is also interesting to note that prior to this date, on the 7th, the SOHO satellite images picked up our old friend the **SunCruiser** as seen and analyzed by Liz Edwards .

On the 15th China experienced a 6.6 earthquake, while on the same day a few very strange radar rings were observed, (shields going up?); And then on the 17th this quote was posted:

> **Hi Kent, This latest round of flares is getting scary...(see the graph for today) @ 5:30 P.M. 09.17.2000 Pacific Time...these Speeds are NOT seen in this whole solar cycle except 1 or 2 times, that's the rarity.**

From this point on events seem to escalate and we attempt to describe and offer the full references wherever possible. Bear in mind solar and magnetometer chart readings are regularly updated, so one will have to refer back to the relevant dates on each chart site.

On the 14th and 15th anomalous objects were observed moving on the LASCO satellite images as can be seen (includes all the relevant commentary) here;

Comment on this object/s on the 15th was as follows;

> **Trajectory is 15 degrees off elliptic.**
>
> **Its aspect ratio is changing as evidenced by differing levels of reflected light coming off of it from frame to frame. Possible rotation of a non-spherical body?**
>
> **Speed across star field seems high. This would tend to make me feel its closer to earth rather that far away, big and very fast.**
>
> **If it's as far away as Mercury's orbit it is very big (remember Mercury transit a month or so ago, it wasn't much bigger).**
>
> **Lack of reflective wings makes me feel it's on this side of the sun, not the far side.**

Also on the 15th the following comment from Charlie Plyler of Elfrad:

> **From: cplyler@elfrad.com (Charlie Plyler)**
>
> **Very strange indeed! There is also an unexplained signal, which corresponds with that time period. At 1300 UT on the 14, the Elfrad detectors began recording electromagnetic pulses in the frequency range of 2.76 to 2.95 hertz. These signals became stronger during the day reaching a peak amplitude at 1730 UT then gradually became weaker and faded out around 2400 UT. The beginning frequency started at 2.76hz and reached its peak of 2.95hz at the same time as the peak amplitude.**

This is an indication of Doppler shift perhaps caused by the earth's rotation if the source is in the direction of the Sun. The peaks also coincided with the time of local noon considering daylight savings time.

Elfrad also recorded the same sequence today, Sept. 15, with the same Doppler shift and same peak frequency times. I have enclosed two screen shots of this signal. The first shows the six hour time period from 1200 to 1800 UT and the second shows the period from 1800 to 2400 UT. The total amplitude today compared to yesterday indicates a slightly stronger signal which may mean the source is closer than yesterday.

.

Various theories as to what this object may have been were posted, was it a comet (but no tail) or an asteroid? No satisfactory answer really came forward. However, it did seem "timely" that the BBC news website decided to place an article up on the 16th titled, ***Asteroids Threaten Earth***, which may be read here;

The threat from space is not science fiction

By BBC News Online science editor Dr. David Whitehouse

Experts are to tell the UK Government that more must be done to defend the Earth from the threat of rocks crashing on to the planet from outer space.

A task force established to assess the threat of the Earth being struck by so-called Near Earth Objects (Neo's) will say that the threat from space is not science fiction but something that should be taken seriously.

Its report, being released on Monday, has been written by Dr. Harry Atkinson, who has worked for several space agencies and government research councils, Sir Crispin Tickell, the former British ambassador to the UN, and Professor David Williams, of University College London.

The publication of their report will coincide with asteroid 2000 RD53 passing the Earth at a distance that is just 12 times further away than the Moon.

The space rock is estimated to be 300-400 metres (980-1,300 ft) across and, although there is no danger of a collision with the Earth, it will highlight the task force's conclusions.

Hiroshima detonations

Astronomers calculated recently that there are nearly 1,000 asteroids that cross the Earth's path which are at The dinosaurs were probably wiped out by the impact of a 10-km (7-mile) wide object that struck the Earth 65 million years ago.

No-one has ever been killed, or even injured, by an asteroid (although there are some suggestions of a deadly medieval Chinese impact), but the consequences of being struck by a rock from space could be extremely serious.

It is estimated that a 100-metre wide object crashes on to our planet every 10,000 years with the force of a 100-megaton bomb.

Every 100,000 years, a one-kilometer-wide object strikes the Earth with the force of 10 million Hiroshima detonations.

International effort

The task force publishing its report was established in January by the government's science minister, Lord Sainsbury, with a brief to assess the danger from asteroids, what could be done to prevent impacts and how the dangers should be communicated to the public.

The report will make 14 recommendations. These include a plea that Britain should:

- take a leading role in a greater international effort;

- improve its ability to detect any incoming objects;
- assess the risks and consequences of an impact;
- implement measures to mitigate any future impacts and new national and international arrangements to cope with the many issues raised.

The need to take seriously the risk to Earth from space rocks has also been echoed by Jonathan Tate of pressure group Spaceguard UK.

On the 25th September, much discussion was banded about as to the possibility of the inbound object being that of Asteroid 1988 OX4, however, and this is where it now gets extremely interesting, on September 29th a rather highly unusual anomaly was observed, initially bathed in a large CME emitting at about 8 o' clock position from the sun. The relevant satellite images, comments and animations were compiled on the internet at the time under the title ***A Dark Entity Defined Within the Solar Plasma***.

Including the following comment:

Date: 10/1/00 7:47:23 AM Pacific Daylight Time

Twenty years ago I was reading excellent book written by some American astronomer (forgot his name) describing instant space travel by using a black hole, based on Einstein-Rosen's bridge theory (from here all space travel wormhole stories originate). This theory implies that on the exit side should be something opposite to the black hole, a white hole. A Sun.

That what Ambilac group predicted could well be, not the Sun birth of a Planet, but arrival of king-size ships not much unlike Death Star from Light Works.

All the best.

IZAKOVIC

So, was this anomaly we observed in fact an interdimensional alignment? We strongly believe that is the case. Thus, one may conclude that if we were to calculate the alignment from where the earth was in relation to this event, and further calculate where we will be in that line of fire, this may give the date in real time, our time line, when this event will occur.

We do understand the alignment and energy transference process that was implemented from within our own dimension on the above dates, however, it is important to note that not all of our known physics is less than correct. For in order for the Sun to give birth to the new planet, here in our dimension, there must exist an equal opposite in existence in our parallel dimension. This is the reason the Suncruiser was on site shortly following the alignment from within this dimension. The Suncruiser is of Eridean origin, and manned by the Erideans, who are true masters when it comes to the manipulation of energies.

For additional information on the Erideans, read our chapter in Volume 1 entitled *THE ERIDEANS*

Once the new planet in the parallel dimension stabilizes in its orbital plane, this will allow and justify the birth of our new planet in this dimension. This process will also help to stabilize our **SUN**.

Earthquake activity:

Three earthquakes occurred during October 1st and 2nd, as anticipated. If you recall, our original prediction was the following:

- Las Vegas, Nevada approx 37N 117W
- Murmansk, Russia approx 68N 35E
- Durban, South Africa approx 30S 30E

On September 30, an earthquake occurred on California-Nevada border, of unknown magnitude, but as there were so many unknowns in the data we felt it was not the Nevada event.

However, a series of three earthquakes did occur starting on the afternoon of October 1st and continued into the early hours of October 2nd. These were:

- Oct 1 14:43 UTC 11.23N 86.69W 33 km 4.6Mb
  Near coast of Nicaragua
- Oct 1 15:05 UTC 68.86N 16.28W 10 km 4.8Mb
  Iceland Region
- Oct 2 02:25 UTC 7.96S 30.53E 10 km 6.8Mb
  Lake Tanganyika Region

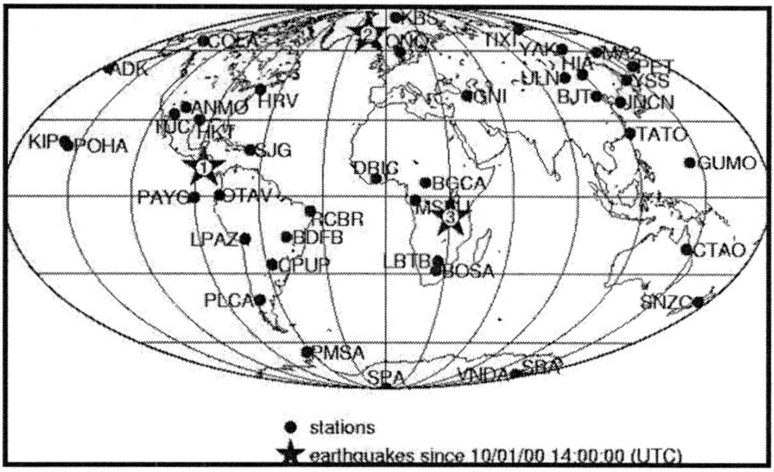

Illustration 13-1 Earthquake map

The main earthquake data was accessed at the following usgs government site at:

http://gldss7.cr.usgs.gov/neis/bulletin/bulletin.html

The deviation of degrees Latitude and Longitude from our original prediction may have averted a further quake, or tsunami, in Sydney. It certainly seems so at this stage, as a series of 3 earthquakes in a fairly tight triangle North and East of Sydney occurred on October 4. These were South of Kermadec Islands, Tonga Islands, and Vanuatu Islands, ranging from 5.1Mb to 6.8Ms.

Further, the deviations indicate that there will be a further shifting of Earth, approximating 24 degrees, after an initial shift of approximately 3 1/2 degrees. You can see this by mapping both sets of Latitudes and Longitudes and noting where the deviations have occurred.

WORMWOOD

The only remaining variable, yet to be accounted for, is the Comet we refer to as **Wormwood**. Our sources have informed us that this comet is presently on a collision course with our planet Earth. We have also been informed that the COMET will make contact with our grid very shortly. Unfortunately, the incoming angle is unknown; therefore we have no way of diverting it thus preventing it from making contact with our Earth Grid.

However, what we can tell you is that the Comet itself will be deflected (or Bounced) once it comes into contact with our grid. However, everyone on this planet will feel this impact and it is at that point they will realize that the material we have touched on prior to this is in fact real!

The energy (Kinetic) that will be created by the Comet at the time of impact will be absorbed by and equally redistributed within the Grid. The effects this will produce on the physical Earth is not known at this time. But above all **do not panic**. Under normal circumstances the impact would be of sufficient strength to knocked earth off of our orbital path. This scenario however, will not be allowed to occur, as preparations have been taken that will hold us in our present orbit, for that is HIS promise.

Those with eyes shall see

Those with
ears shall

hear

Chapter 14

THE AVAILABLE EVIDENCE

On September 9th 2000, the Solar Cruiser was visible in one of the Lasco satellite images, as captured and enhanced by two dedicated people that regularly carry out such data capturing. [1]

Figure 14.1
Solar satellite image September 9th showing sun cruiser

As a further explanation we now review the many events that have occurred since that time.

On September 12th an exceptional CME was observed from the sun and dubbed the Madonna Object

Satellite frame capture below:

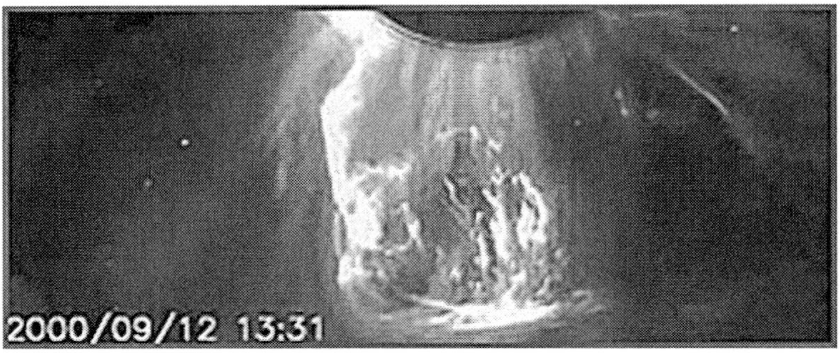

Figure 14.2 Solar Madonna object

The following comments on this event were posted on the Internet via science at NASA.

> Sept. 13, 2000 — On Monday, Sept. 11, the sunspot number dropped to its lowest value of the year. The face of the Sun looked remarkably blank with only a few tiny spots visible in telescopes. It was a remarkable sight considering that the Sun is well into the maximum phase of its 11-year sunspot cycle. The Sun has been almost constantly peppered with spots since "Solar Max" began earlier this year.
>
> Yesterday, just when it seemed that a brief interlude in solar activity was at hand, the Sun unleashed a brilliant full-halo coronal mass ejection (CME)

While from the space Discovery news briefs we have extracted this relative portion.

Sept. 15, 2000 — A solar shock wave that hit the Earth early Friday is either a dud, or just revving up.

The brunt of the great spotless flare that erupted from the sun on Tuesday was expected to reach Earth late Thursday, but so far the blast of super-hot particles has been less than remarkable.

But scientists monitoring the solar blast have detected changes in its magnetic field that could cause trouble over the next few days.

When the shock wave first hit around 1 a.m. Friday, only a rather weak rise in the solar wind was detected by spacecraft — not enough to cause a geomagnetic storm, said space scientist Kent Doggett, a forecaster at the Space Environment Center in Boulder, Colo.

The magnetic field of the shock wave was initially oriented opposite the Earth's magnetic field, and since opposites attract, there wasn't a powerful repulsion to release a lot of energy, reducing the impact of the solar punch.

But by Friday afternoon, space scientists said the shock wave had switched its magnetism and was repelling against the Earth's field, increasing the possibility of storm activity.

Solar scientists will keep an eye on the shock wave over the weekend in case it ramps up suddenly and causes trouble for satellites, Doggett says.

Although spotless flares throw a kink in the process of trying to forecast space weather by monitoring sunspots, they can be anticipated by watching for long stringy filaments of hydrogen on the face of the sun, says solar physicist David Speich, also of the Space Environment Center.

"A filament is a clump of solar mass — hydrogen — that sits on the sun," said Speich. Sometimes those filaments break free of the magnetic fields that hold them in place, sending all that mass into space. That's what happened Tuesday when a filament about fifty times as long as Earth's diameter was seen across the face of the sun, then disappeared.

"This filament disruption is a fairly rare event," said Speich. "It's a tough forecast."

At the same time, solar observing satellites picked up a burst of energy from the sun, including a "full halo" coronal mass ejection that usually indicates that the flare was aimed right at Earth. It's similar to looking straight down the barrel of a gun and seeing the puff of smoke before the bullet hits.

While a comment added at 06:05 UTC on September 15: A solar wind shock was observed at ACE at 04:05 UTC. The interplanetary magnetic field has since then been northwards and only a modest increase in geomagnetic activity has occurred so far. Should the IMF swing southwards active to major storm conditions may occur.

A large coronal mass ejection, probably a full halo CME, was observed following a very long duration M1.0 event which peaked at 12:13 UTC. The source of this event was a huge filament eruption in the southern hemisphere, stretching about 30 degrees west from region 9158 (and across the central meridian). This CME could cause a serious disturbance on September 14 or 15 as it impacts Earth, minor to severe storming

Spetember 23rd saw the return of the solar cruiser as seen on the Lasco 2 satellite frame of 2000/9/23 23:50.

Below is a detail of the cruiser from that frame:

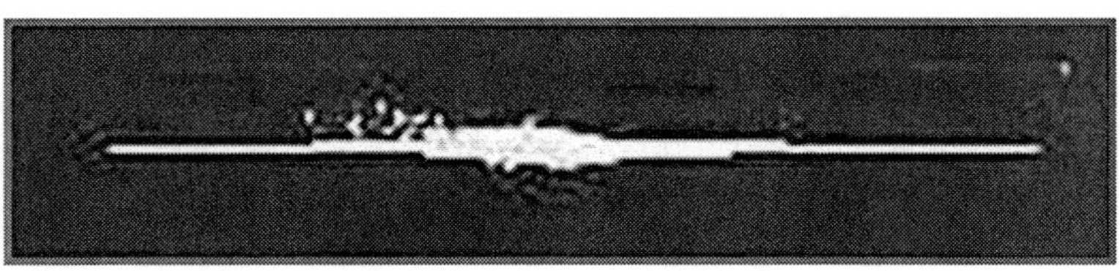

Figure 14.3 Enlargement of Solar cruiser

Dark Entity

One of the most significant events was that of the solar anomaly of September 29th, as observed and commented in detail at Cyberspaceorbit web site, where we place just a few comments below:

> Any idea what the object in? It almost looks as if the sun is reflecting off its right side.
>
> MESSAGE BOARD: 04:49:56 PM 09/30/00
>
> Well, it looks like an object (the circular thing), but...
>
> A. Size is -huge- as it appears close to the sun (being enveloped in the CME discharge)
>
> B. There is not enough light reflection for me to believe it is totally solid.
>
> So, it's either a large flying saucer or it's Nibiru....
>
> Or, it's a much smaller object that has come from behind Sol with it's own attendant gas cloud or debris field.

Many likened this scenario with the following news section from the Washington Post:

Mystery Heavenly Body Discovered, a front page story - 31-Dec-1983

> A heavenly body possibly as large as the giant planet Jupiter and possibly so close to Earth that it would be part of this solar system has been found in the direction of the constellation Orion by an orbiting telescope aboard the U.S. infrared astronomical satellite. So mysterious is the object that astronomers do not know if it is a planet, a giant comet, a nearby "protostar" that never got hot enough to become a star, a distant galaxy so young that it is still in the process of forming its

first stars or a galaxy so shrouded in dust that none of the light cast by its stars ever gets through. "All I can tell you is that we don't know what it is," Dr. Gerry Neugebauer, IRAS chief scientist for California's Jet Propulsion Laboratory and director of the Palomar Observatory for the California Institute of Technology said in an interview.

The most fascinating explanation of this mystery body, which is so cold it casts no light and has never been seen by optical telescopes on Earth or in space, is that it is a giant gaseous planet, as large as Jupiter and as close to Earth as 50 billion miles. While that may seem like a great distance in earthbound terms, it is a stone's throw in cosmological terms, so close in fact that it would be the nearest heavenly body to Earth beyond the outermost planet Pluto. "If it is really that close, it would be a part of our solar system," said Dr. James Houck of Cornell University's Center for Radio Physics and Space Research and a member of the IRAS science team. "If it is that close, I don't know how the world's planetary scientists would even begin to classify it."

The mystery body was seen twice by the infrared satellite as it scanned the northern sky from last January to November, when the satellite ran out of the supercold helium that allowed its telescope to see the coldest bodies in the heavens. The second observation took place six months after the first and suggested the mystery body had not moved from its spot in the sky near the western edge of the constellation Orion in that time. "This suggests it's not a comet because a comet would not be as large as the one we've observed and a comet would probably have moved," Houck said. "A planet may have moved if it were as close as 50 billion miles but it could still be a more distant planet and not have moved in six months time.

Whatever it is, Houck said, the mystery body is so cold its temperature is no more than 40 degrees above "absolute" zero, which is 459 degrees Fahrenheit below zero. The telescope aboard IRAS is cooled so low and is so sensitive it can "see" objects in the heavens that are only 20 degrees above absolute zero. When IRAS scientists first saw the mystery body and calculated that it could be as close as 50 billion miles, there was some speculation that it might be moving toward Earth. "It's not incoming mail," Cal Tech's Neugebauer said. "I want to douse that idea with as much cold water as I can."

During the next few days, there was much activity and comments on asteroid and cometary activity, especially as many reports of new discoveries of comets were being made as notified in NEAT (Near Earth Asteroid Tracking) at the NASA site, of which we place the following comment:

12P.10] Results of the JPL Near-Earth Asteroid Tracking (NEAT) Program

E. F. Helin, K. J. Lawrence, S. H. Pravdo, D. L. Rabinowitz (Jet Propulsion Laboratory), NEAT Team

The Near-Earth Asteroid Tracking (NEAT) program is the first fully automated system for controlling a remote telescope, acquiring wide-field digital images, and detecting Near- Earth Objects (NEOs). Under an agreement between the Jet Propulsion Laboratory (JPL) of the California Institute of Technology and the U. S. Air Force, JPL is provided access to a 1.0 m telescope located at the 3000 m summit of Haleakala crater in Maui, Hawaii. By scripting 20 sec exposures every 45 secs, this system can image up to 800 sq. deg. thrice nightly to magnitude limit V=19.5, and can detect asteroids with 90% efficiency to V=18. As of July 1998, NEAT has surveyed approx. 60,000 sq. deg., detected more than 25,000 asteroid, and discovered 33 new NEAs, 2 comets and unique object 1996 PW. The current rate of NEO detections, including incidental redetections, is about 1.5 per 1000 sq. deg., of which half are larger than 1 km in diameter. NEAT leads the field in the detection of these largest and most hazardous NEOs, and also in the detection of Aten asteroids with orbital periods shorter than 1 year. A proposal is currently under review to run the NEAT program on three Air Force telescopes, 18 nights per month. Such a system would cover the whole night sky 3 times per month to V=20, and detect 90% of the NEOs larger than 1 km in 20 years. Simple improvements, such as operating the CCD at a lower temperature and improving telescope tracking, would lead to 90% detection in 10 years.

On October 2nd we issued this posting at Orbit

> From the time the alignment begins (with our sun) we will begin to experience a sharp steady increase in alterations to the mag. poles (Earth's). I estimate to see about a 3 and 1/2 degree shift between now and Dec.13-20th "this year".
> On the Dec. 21st (birth event) there will be a drastic change in the mag. poles shifting, at that time it will be a major shift, and this shift at this time will occur instantaneously, verses the 3 1/2 deg. much slower shift (time wise). I am expecting that sudden shift will be in the neighbourhood of some 24+ deg.

We were referring to the three quakes that we predicted many weeks previously, that in actual fact came about, but all 20 degrees out?!

During this period, and since, we have experienced much solar activity, including large CME (Coronal Mass Ejection) activity, resulting in global auroral displays, such as the one over Toronto in Canada....we quote;

Date: 10/4/2000 10:23:00 PM Pacific Daylight Time

> Location: 20 km north of Toronto Ontario;; i++)
> It is now Oct 5, 01:15 Eastern (06:15UT). Massive Auroral display. Horizon to horizon to 90deg Azimuth. Persistent. Duration, unknown. 30min + at least. (Just came up from the basement office). Almost 100% sky coverage at times. I have seen aurora's many time before and this is by far the biggest and most spectacular I have EVER seen at ANY latitude.

Date: 10/5/2000 5:44:10 AM Pacific Daylight Time

The magnetic storm that began on Tuesday continues and has become more intense today. Planetary magnetism readings as high as "7" have been reported and a magnetic-A index of greater than 50 was reported this morning ... making this a major to severe magnetic storm event.

While on October 6th, we received this earthquake bulleting affecting Japan;

00/10/06 04:30:19 UTC 35.56N 133.05E Depth: 10.0 km 6.7Ms WESTERN HONSHU, JAPAN

From October 7th events seemed to pick up and we were receiving much data from the Elfrad group, and the solar reading stations.

A comment by one known as "Anony mous" was posted on October 7th regarding the incoming we talked about previously.

10/7/2000 9:30:45 AM Pacific Daylight Time

> ...it is coming from that quadrant, about 6 days out. Impact 10/13/00 2 PM GMT over Mermansk/Archangel area at an angle of 31 degrees positive from True Horizontal. It will skip off, but the seismograph and the quakes will go wild.] ANONY MOUS

On October 8th, we observed an interesting phenomena on the Lasco 2 solar satellite, that of TWO solar cruisers practically side by side.

Then on October 11th we witnessed what was termed the "October Solar Torpedo", whereby an "incident" was observed around the sun.

10/10/00 9:03:39 PM Pacific Daylight Time

> Kent: See attached frames. Comet? Something else...its moving quickly as evidenced by its change in location between frames. Also note several ORCA effects in last 8 hours of C3 frames—9:08:10 PM: Just found out,,
> "NASA is saying its a new Kuertz

An "anomaly" travelling at rapid speed was seen to head direct into the sun, while the reaction of the sun may be observed on the satellite image 10/11/00 07:27

However, just prior to "impact" NASA pulled the feed, as commented on below;

> 20001011_0727 pulled feed 20001011_1006— Okay that does it, clue, the damned object was artificial! HAPPENED THE SAME WAY ON LAST TORPEDO! After the 1998 TORPEDO they lost the whole freakin spacecraft—not merely glitched but gone away, remember that? Can anyone think of a phrase that rhymes with "Lockheed lizards?" I am trying, TRYING to be poetic and...nice!

Comment

> Noticed C2 is current but C3 out since 8:00UT, Look at last Navy C3 whole left side blocked out and feeds end there for now. live and current raw data, has been replaced with frame, 10/10 02:42, captured by the SOHO cameras the day before the torpedo incident. This data-intact C3 image is MOST ANOMALOUS!]

On October 13 an earthquake occurred on the Atlantic ridge, as can be analysed here;
While on the same day we issued this statement (read at)

Date: 10/13/00 10:32:51 PM Pacific Daylight Time

> Kent: check out soho images 18:42 [detail-left] and 19:42....if that is it and sure looks like it Wormwood entered time frame at 18:42...must of then taken it's bearings and left on intercept course at 19:42.... It's about 12:30 here right now...I hope that I'm wrong but looks like we will be in the wrong place at the wrong time within a couple hours...like I said I hope I'm wrong...JIM, AMBILAC

10/12/00 6:19:36 PM Pacific Daylight Time cplyler@elfrad.com writes:

> A high amplitude magnetic disturbance was detected by Elfrad at 00:31 UT, 10/13/2000. An extremely high magnitude pulse was recorded at 00:51:58 UT. The signature is unusual for a magnetic storm. Very energetic.. Charlie Plyler at Elfrad

While...

Date: 10/13/00 9:31:31 PM Pacific Daylight Time

Kent, 10/13/00 2:00 PM GMT came and went without incident. "Intel reports intruder fragmented and lost punch. Large piece skipped off the grid over the Central Mediterranean in the area of Malta (North and West of Cairo)
Grid is back on line and powere. One disturbing note is that sources in the Azores report a strong electromagnetic residual signature coming from the Brookhaven area. Hypothesis is that an electromagnetic beacon is DRAWING comets and asteroids in our direction. It shuts down so that objects will "Just Miss" us. Their calculations were off this last time for sure. Someone out there likes us and fragmented the beastie or we would have been hit a glancing blow. The grid would have stayed up, but would have been weakened considerably. The consequences of this signature in the crust resulted in shock waves that may have ignited that mid Atlantic quake. We are still running calculations to confirm this. Anony Mous

While the following day another comment from the same source;

Date: 10/14/00 10:01:04 PM Pacific Daylight Time

Three came in on our Mother Earth, One in Oklahoma, One managed to stay on course (also a small one like Bolide) and a pretty sizable one bounced off the Grid a little North and East of Cairo. To the best of our knowledge the Bolide was the only one near enough to "Put on a Show". The Grid is firing up again and will stay at peak through December 20th or so. As Jim at Ambilac says, "It may be a Wrong Place, Right Time Scenerio", We believe however that a somewhat Happy Ending is in the Offing. Anony Mous

Fires in the Sky:

It was during this period that many witnesses came forward to report the observations of unusual fireball and strange lights in the skies.

Witnesses see fire in the sky WICHITA, Kansas, October 13 – Kansas watches as a strange sight lights up the night, as reported on the local ksnw news site

.

Witnesses see fire in the sky

> WICHITA, Kansas, October 13 – Kansas watches as a strange sight lights up the night.
>
> "THERE'S ONE BIG WHITE LIGHT and around it was like five little balls, just lights, just following them. And then thru the sky there was just like vapors and vapours were like different colours – blues and oranges.
>
> It was just odd," said eyewitness Anthony Davis of Valley Center. We've had reports of the fire in the sky from all over the Midwest. From McCook, Nebraska in the northwest, through Great Bend to El Dorado. In Wichita, hundreds saw the sight.Gary Bishop of south Wichita said, "So I looked over my shoulder and right across the sky from the southwest to the northwest, down a little bit. There was a meteor that left a trail of blue lights.It was beautiful."
>
> As the object screamed across the sky, 911 couldn't keep up with all the calls that were coming in. 911 supervisors say they had more phones ringing than dispatchers to answer them. They say they had around 150 calls from witnesses saying they saw everything from space aliens to a plane crash.
>
> While dispatchers were getting all those calls on the meteor shower, they did have a serious car accident to deal with as well.

Luckily, supervisors say that call got through without much of a wait and response time wasn't affected. So what was it in the sky? We talked to the North American Air Defense Command to find out. Master Sergeant Larry Lincoln told us, "It seems to be a very heavy meteor shower. Our weather people told us to expect something like this from the 10th to the 29th of October, with the heaviest coming down around the 15th. It seems to probably be coming down a little early. The mission of NORAD U.S. Space Command is tracking.

An article appearing on the NASA science News web site blames the fiery display on the 4th stage of a Russian Proton rocket launched earlier in the day from Kazakstan, Russia.

The rocket was used to launch 3 satellites for the Russian Glonass global positioning system array.

The casing was "3.4m in diameter and 4m long that weighs some 800kg," according to Alan Pickup, a satellite decay expert who works at the United Kingdom's Astronomy Technology Centre at the Royal Observatory in Edinburgh.

And a comment from a local observer..

Date: 10/14/00 12:48:53 AM Pacific Daylight Time

I live in the Okla City metro area. Tonite on the local news, they reported that people all over the state had been calling in about something they had seen in the sky. The news station reported it as a bolide type meteor. Apparently it crossed a large portion of the state from South to North. When it was first seen, people noticed a 'smoky' or dirty tail (late evening), but after the sunset it had a tail similar to a comet. In its final stages it apparently broke into several pieces. Unfortunately we didn't see it here because we are having cloud cover for the first time since June.

It may be of interest to note the following military activity while the above events were occurring.

Date: 10/11/00 2:17:14 PM Pacific Daylight Time
Louisiana National Guard unit has been called up to report to Saudi Arabia to babysit an air force base. But, assignments can be redone in a minute, once the unit is assembled

Date: 10/11/00 3:38:16 PM Pacific Daylight Time
I an interesting article about the movement of 44 M1A1 Abrams battle tanks from Ft. Hood Texas to four locations in central Minnesota. That's a pretty nicely equipped National Guard unit! —ex-NSA analyst

Monday October 9 11:25 AM ET U.S.,

NATO Forces to Train Off Atlantic Coast

> MIAMI (Reuters) - More than 30,000 military personnel from the United States and its NATO (news - web sites) allies will conduct joint training exercises off the eastern United States coast and the Caribbean this month, a military spokesman said on Monday.

The air, ground and sea exercises will begin this week and run through Oct. 28, primarily off the coast of Virginia, North Carolina and in the Caribbean, said Lt. Cmdr. John Kirby, a spokesman in Norfolk, Virginia, for the U.S. Second Fleet and NATO's Striking Fleet Atlantic. The operation will combine two exercises — NATO's "Unified Spirit" exercises, which are conducted every four years among the North Atlantic Treaty Organization (NATO) members, and the U.S. Joint Task Forces exercises held every five months among all the U.S. military branches to prepare for "demanding scenarios and challenging conditions they might face when they deploy overseas," Kirby said.

> Date: 10/12/00 1:25:51 PM Pacific Daylight Time
> I noticed last week or the week before (between 25-Sep and 5-Oct, sorry for not having exact dates) that military hardware was being moved by rail in the Syracuse, NY area. I live near the train yards in Minoa (Syracuse suburb) and the yard was full of M-60 tanks on

flatbed rail cars. A few days later, I noticed a train moving on the rails adjacent to Onondaga Lake full of HUM-Vs and Jeeps. I couldn't make out the unit designations, sorry. Actually, it didn't hit me until I saw the 2nd train, then the li'ght bulb started to glow. For what it's worth .

Date: 10/13/00 9:04:33 AM Pacific Daylight Time
convoy on U.S. highway 321, north Carolina...heading north.

Hi Everyone,

The skies in mid-coast Maine have been filled with,(I assume-I'm not an expert on military aircraft), huge military transport planes. This is very unusual, and I can't state this too strongly. They are passing in a NNE pattern, heading off the coast and I presume to Gander, Greenland for refuelling and then on to Europe. Brunswick Naval Air Station in about 60 miles south of me; I don't know if they are originating from there. They are flying very low, almost lumbering in their flight. I saw one at 10:30, another one at approximately 1200 and then 2 very close together at about 2pm. Now I know all of this isn't any surprise with all that's happening in the Mid East, but having it unfold before my eyes is scary

Date: 10/15/00 10:22:07 PM Pacific Daylight Time
hi kent...military plane flying Very low near Newton, nc. upper portion of plane was dark green, lower and belly white. US MARINES in all caps near door. the plane had a long thick antenna type thing sticking out of the nose. AF was on the tail. oddly, the plane was landing! there is no airport big enough to land even a Learjet here! where did it land? the plane appeared to be attempting to line up with a GWEN tower. the plane was heading nne. there were some type of tanks on the wings and a large tank in the belly. also, there have been no chemtrails whatsoever in the past 6 days (highly unusual). suddenly today they went wild spraying. webs were flying everywhere. they looked like long spider's webs but they were blue. many people around nc are reporting appache sightings, fully armed and flying low over

people's homes. mostly out in the countryside. even with all of this going on...i have a strong hunch that things are NOT going according to plan. just an old lady's hunch mind you. (4)

Next, came the start of what will, in our opinion, be the Middle East crisis.

Quote:

As we are contemplating the possibility of the Arabs destroying their own Dome of the Rock, we receive the shocking news that the Israeli Government is allowing the Temple Mount Faithful to carry out a ceremony to lay the cornerstone of the Third Temple, under the protection of Israeli military! Can these two events happen in conjunction with one another?

The Biblical prophecy Armageddon cannot happen until the Third Temple is rebuilt — or in the process of being rebuilt — "The marble corner-stone... weighing four and a half tonne, will be carried on a flat-bed truck, covered with Israeli flags..." — Laying this cornerstone opens the door to ARMAGEDDON

http://www.rumormillnews.com

Meanwhile, back at the "Ranch"

Date: 10/16/00 3:23:31 AM Pacific Daylight Time
From: cplyler@elfrad.com (Charlie Plyler)

Frank, I think you need to see this. Please check your recordings for 2:19 UT on 10/14. This is around the time of the fireball. The really interesting item of note is the signature. According to our detectors, the event was caused by a non earth based anomaly. The array we

use for wave cancellation detected a possible scalar pair, wave form. This was recordable for around three seconds before the main spike. See pic. Top view..North-South array Bottom view...from same array however using wave cancellation technique.
Whatever the cause, it was very powerful. Andrei may be interested in this event as it may be non meteoric in origin.

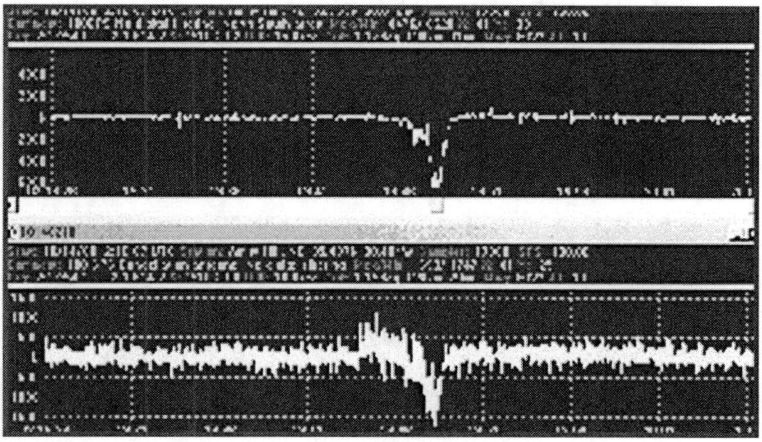

Figure 14.4 Elfrad graph

On October 17th Frank Condon of Geo Seismic labs replied with this comment:

Date: 10/17/00 2:11:34 AM Pacific Daylight Time

Here's what I detected from Southern California at the approximate time you specified. However, the anomaly shows up 78.2 seconds earlier than what you recorded from the east coast. Apparently the objects in question passed by here that much sooner! Could we figure the airspeed of the bogie with the timing information? Please let me know what you think about this aspect.
Regards, Frank Condon, Geo-Seismic Labs

The relevant read out is illustrated below.

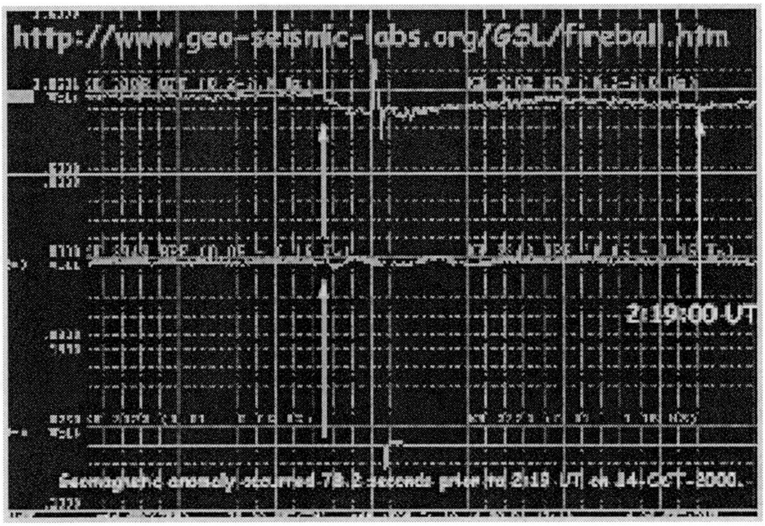

Figure 14.5 Geo seismic lab graph

While the day previously we witnessed this intriguing solar satellite picture; http://www.cyberspaceorbit.com/frontpage/10-16c2.gif

Figure 14.6 Solar satellite image 16th September 2000

While below is the accompanying Proton flux chart.

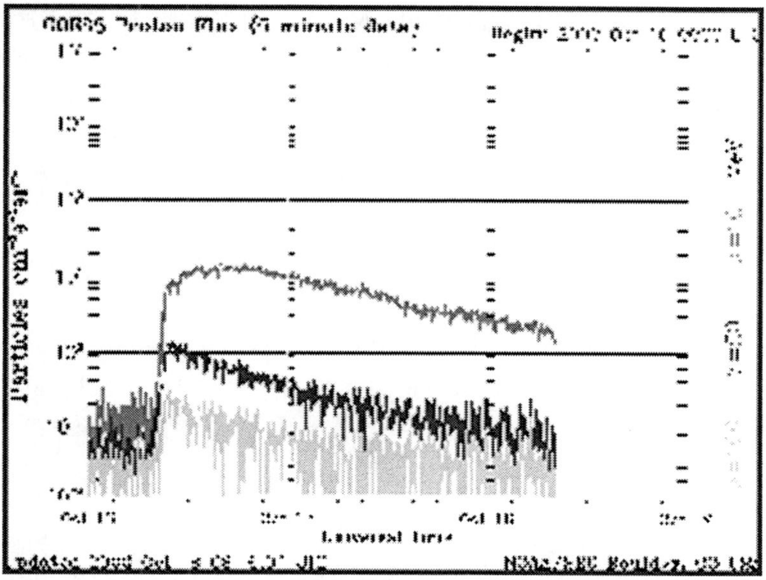

Figure 14.7 Proton Flux chart

During the middle and latter part of October, the attention was turned to the incredible number of fireballs and associated multicoloured lights observed in the skies over USA, Australia and parts of Europe.

While on October 27th NASA CUTS US OFF. DID ANOTHER PROJECTILE PIERCE THE SUN?

Date: 10/28/00 2:54:26 AM Pacific Daylight Time

In a message dated 10/27/00 6:34:13 PM Pacific Daylight Time, ANON writes:

<< If you look and see the Streak (comet..asteroid?!?) moving from Left to Right...Now The Sat. pictures went all haywire after these?!!?!? >>

The reference satellite image people were referring to (LASCO) was dated October 27 at 01.06, and the file reference was File No 43185134.ft

In addition the following comments were made at the time

> 20001011_0727 pulled feed 20001011_1006— Okay that does it, clue, the damned object was artificial! HAPPENED THE SAME WAY ON LAST TORPEDO! After the 1998 torpedo they lost the whole freakin spacecraft—not merely glitched but gone away, remember that? Can anyone think of a phrase that rhymes with "Lockheed lizards?" I am trying, TRYING to be poetic and….nice!

November literally entered with a blast, severe weather, high winds and serious flooding in Asia and the UK, which is still on-going. Over the last few weeks we have consistently observed massive Coronal Mass Ejections (CME's) from our sun, bringing with it the accompanying electromagnetic storm bombardments. "Coincidentally, as we compile this article we are literally observing a Massive proton blast.

The associated proton flux chart http://solar.sec.noaa.gov/rt_plots/pro_3d.cgi showed a massive surge at the commencement of this event. As this chart is updated continuously A capture of the reading is seen below.

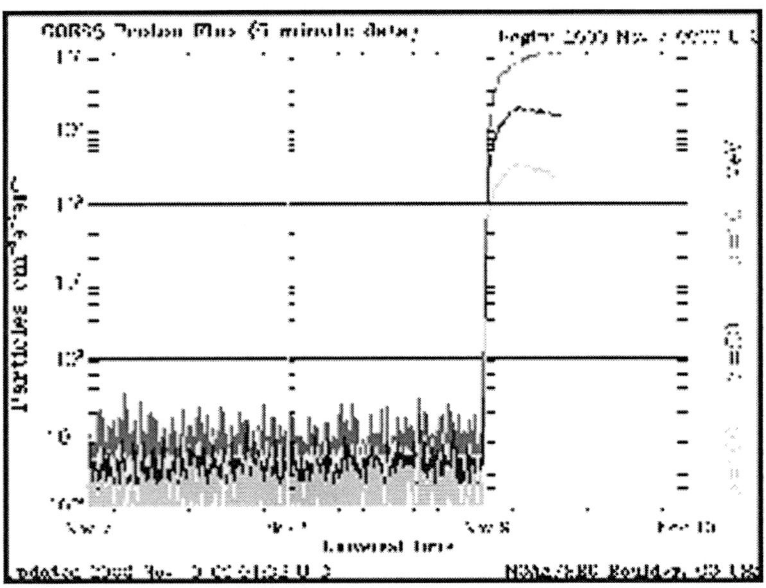

Figure 14.8 Proton flux chart 2

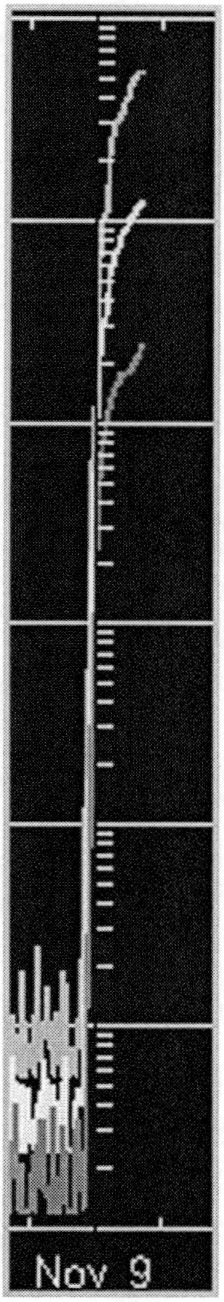

On November 26th multiple solar X-class CME's were witnessed, as can be observed at

http://www.cyberspaceorbit.com/sun/11-24gallery.html

Six to seven CMEs headed their way towards Earth. The most powerful solar flare of the current series erupted near sunspot 9236 at ~1650 UT (1150 EST) on Nov. 26th. The X4-category blast triggered a strong radio blackout over the sunlit side of Earth.

All the above quotes were taken "as is" from Kent Steadman at www.cyberspaceorbit.com

Figure 14.8a
Capture of chart reading

(Endnotes)

[1] Rusty's website
http://www.homestead.com/Solar_Station/enhancements_3.html
Liz at http://www.iwonderproductions.com/sep97.html

Chapter 15
TIME - WHAT DOES IT MEAN?

In the beginning of May 2000, a series of events began which we have all been following very closely. There have been numerous occasions where the Sun has been observed undergoing some rather awesome metamorphosis. To say that this very visual display has held people in suspense would be grossly understated. We have attempted to relay the information, which we have received, and have been instructed to pass on as we are instructed to do. BUT on the way, we learned that although we witnessed the EVENT occur precisely when we were told it would occur we began to notice that the ramifications of the EVENT did not necessarily occur at that time. Knowing that the information we were receiving was 100% accurate that left us to attempt to fathom out the reason why this worked out the way that it does. This was quite an eye opening "learning" experience-, which we will now try to shed some light on for you.

We will be utilising one of our computer models to demonstrate how this is calculated. We have been working with this full model for sometime now, and one of which is first revealed in Vol. 1 of the Giza-Genesis series. The beauty of the overall model is that it CLEARLY establishes the relationships between ALL of the constructed monuments located in and around the Giza Plateau. After learning HOW to read it, you realize that the size and the placement of every Pyramid, Temple, and every procession way is built, and placed, so as to reflect the whole model.

This is the very reason that it has defied any reasonable VALID interpretation, for without the entire model to work with; this would be a near impossible task.

As one will recall all mirroring of the complex was generated from the apex of the pyramid Khufu. In Volume 1 we go into great detail as to establishing the alignments, which in turn dictate the size, as well as the location of each of the various structures located on the Plateau. However in this section we will not be spending much time with those alignments, as they are related to another aspect of that which this article is about. This article is about Earth's relative "0"Time, in relation to it's Galactic "0" Time. There are 3 levels of Time (relative), that any and all matter adheres to, we will use the #1 to represent the Earth (as this is the Time which we are concerned with at this time). To this we will add "0", making our Earth's Time reference read as "10".

As Earth's Time is calculated on a 360 Degree complete orbital path abound our Sun, we must then find our Earth's relationship to the Sun's relative "0" point (reference point). As all things within a perfect system the Sun too has its "0" (Relative), time it functions within. As the Earth derives its energy directly from our Sun, therefore as the Earth is directly influenced by the Sun, which is governed by its relative time, we must then add this into our equation. Keep in mind that the Sun also must adhere to the same system of relative Time that the Earth does. So to our "10"(Earth's relative Time) we then add the "0"Time factor of the Sun and we will now show this as "100".

The same governing criteria apply to the Star, which our Sun is itself orbiting, so once again we must take this Stars' relative time and factor it into its place within our Earth's. And as this is two steps back we then show this as "1000".

The book of Revelation, describes many things but the two most important things to bear in mind are the 4 Horsemen, and the 1000yr.reign. The other important items are the 7 stars, 7 candles and the 7 churches. Next is the number 144,000.

Utilizing our 360 degree mathematical model, we find that from the centre there are 4 equal 90 Degree quadrants. In an earlier section we explained how the 12 houses are derived etc., we had also further explained how there are 3 houses, at ALL times having their maximum influence on our planet in any given quadrant. These 3 houses and the effect on our planet, at any given time dictate our Orbital spin. Because time is constant it cannot exist solely in that quadrant. The Earth continues to spin as the next quadrant enters the space. This process continues until the original quadrant has gone Full Circle.

As our Sun is our second Time line it is only fair to say that 1 Earth year (of Time) is (10x4=) 40 Sun years.

To put this into perspective, most will be aware of the Montauk Project [1]

This is still an on-going project, which was originally developed in the 1940's In addition some will be familiar with the "Philadelphia Experiment", through which it was implemented. Some may, or may not believe in those Time Travel experiments. However, based on our findings we choose to believe that this was and still is technology that is being used. We state this for the following reasons:

Their experience is a result created from affixing their equipment to an off planet time frame "0" Time. This is the reason that they were caught up into their present 40-year Time Loop.That is, from the Eldridge (Philadelphia experiment) in August 1943 to Montauke in August 1983.

This is also substantiated by the fact that they are using 24-25 Dimensional Physics, this being the maximum level they could have obtained (which was given to them by Rogue groups of ET's), given that those who gave them this data possessed hidden agendas of their own.

Again, the same process is held to be consistent when we calculate the Time element of the Sun to its relationship to its source or primary Star. However we must still factor that data into our time reference. This is 10x4x10=400.

Which draws us to these numbers:
360x4=1440
360x40=14400
360x400=144000

Now we will demonstrate how this information is derived from the model.

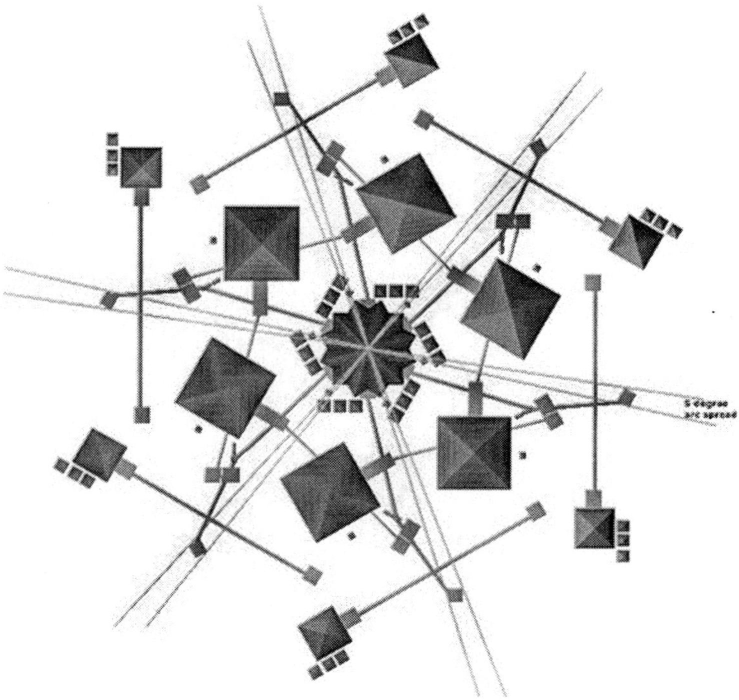

Illustration 15-1 Plateau Time alignment 1

Here you will see that we measure from corner to corner of two of Khufu's Valley Temple locations, 180 Degrees opposite to each other. These lines cross precisely in the CENTER of the Pyramid of Khufu. (If you were to use the apex of Khufu as a pivot point and extend a line out to the very centre of the Valley Temple of Khufu, and were to draw a circle you would notice the following: that this line would not only cross at the precise corner of the Pyramid Menkaure, but it also crosses the precise corner of Menkaure's Mortuary Temple.)

The arc deviation between the two Temples (corner to corner) is 3.142857 Degrees.

The arc spread taken from the very centre apex point of Khufu to the two extreme outside corners of its Valley Temple is exactly five Degrees. This is the very reason the capstone was never placed atop Khufu. In previous articles we went to great lengths to explain the value of "0" to being that of 1.42857Degrees. Also that in a complete system there must be present 7 relative "0's". Total= 7x1.42857=10. OR another way of putting it would be, from its centre it is precisely 5 in and 5 out.

Now, in order to be able to relate this into the Time aspect, it must be done in the following manner. We have four quadrants, each separated by a "0".

If we them were to justify our orbital Time we would proceed this way 4x1.42857=5.7142857.

The Time of Earth's orbit around the Sun is approx. 365.24 day's. Immediately we can eliminate the .24, this due to the fact that we are measuring the Time as it relates to the outer surface.

As for the 5.7142857 this comes about because we are measuring Time as it equates to the 3 levels, those being 10,100 and 1000. Now by moving the decimal point over one it would read as 57.142857. We then can eliminate the 7, as we know this to be equal to "0", (7x1.42857=10). We then would read this to be 50.142857 we then progress and advance our decimal point again and find this to read as 501.42857. We then convert our 1.42857 (knowing it to be the value of "0") so our number then reads as 500.

Remember that is to the centre, on the whole it will be 5+5=10, 50+50=100, 500+500=1000.

There is much that we could say about each of the following diagrams, all in regard to the alignments and the points which in turn give the defining measurements to each structure etc., this however is not necessary at this time. One should however take notice as one view each diagram.

In earlier work we explained that there are three primary star lineages, which govern, and substantiate our existence, the full model itself reflects just that.

The Pyramid located on the Plateau representing Earth is Khufu, the Pyramid of Khefre represents our Sun, leaving Menkaure to represent one of the Stars, from the Suns direct lineage, located within the Pleiades star system.

Corner through corner, passing across the apex of the Pyramid of Khufu, we find this arc spread to be quite precisely 37 Degrees. Please bear in mind that you are viewing these pictures TWO Dimensionally (2D). Also remember that this is a lineage and there are two sides to each lineage.

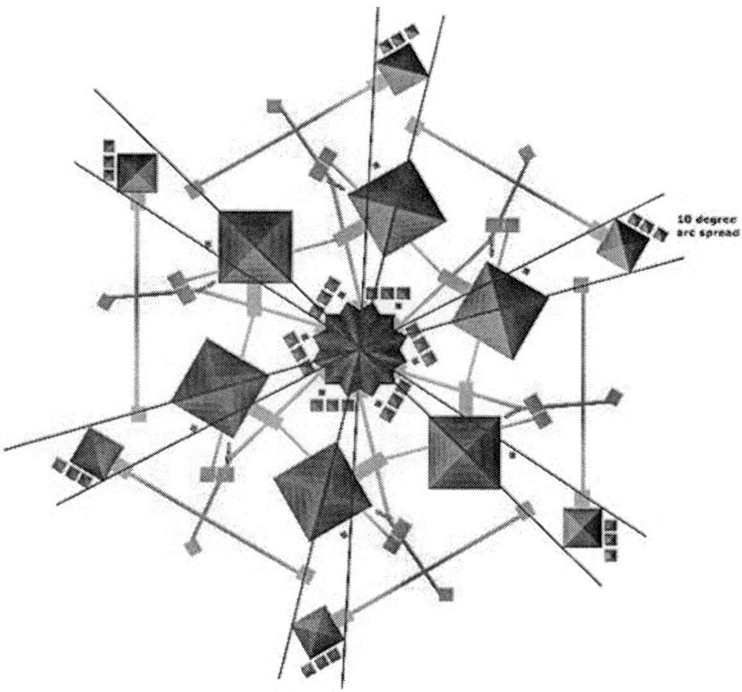

Illustration 15-2 Plateau time alignment 2

Here we have repeated our process using a second lineage, we will call this the Orion lineage. What we are establishing here are the parameters of the four quadrants.

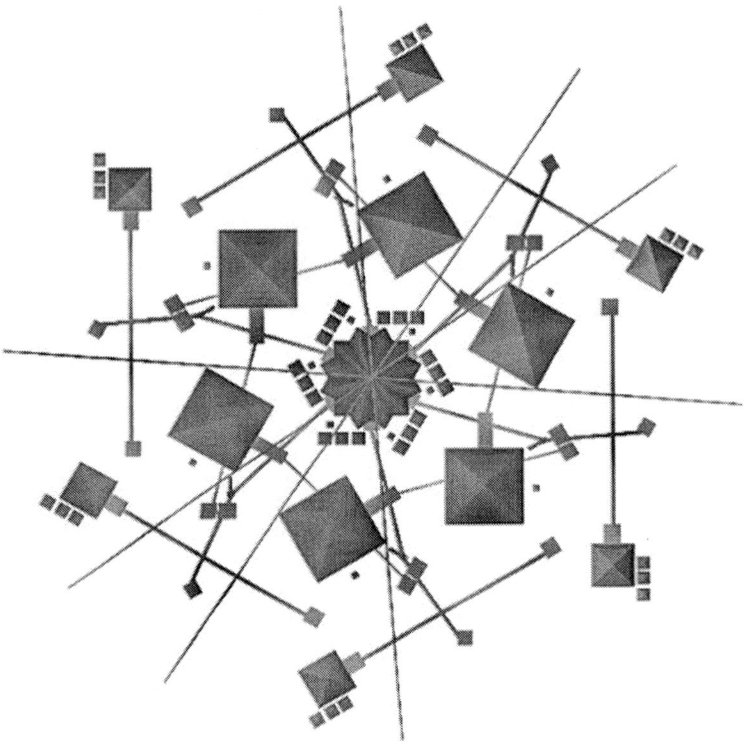

Illustration 15-3 Plateau time alignment 3

In the above picture we are repeating the original process of alignments thru the apex of Khufu, only this time using the Pyramid known as Menkaure. We find that the arc, measured from the apex of Khufu, to be exactly equal to <u>10</u> Degrees.

It is difficult to actually appreciate the full model itself (without the use of the 3D animation) viewed from a 2 dimensional picture, but we will endeavour to demonstrate. BUT remember you are seeing 2 Dimensional imagery.

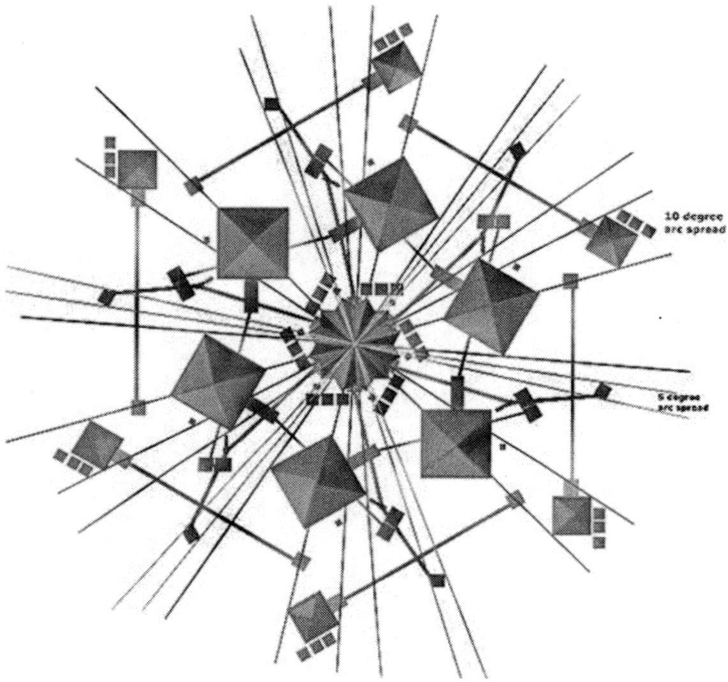

Illustration 15-4 Plateau time alignment 4

In the above diagram we find that the 7 Degree arc spread to be great interest, not only does it reduce the Sun's overall arc to a precise 30 degree's, we find that it defines the element of "0" that we found in the Earth's Time model mentioned previously as 5.7142857. Or the alignment of "0" to its second point. Even more important is the fact that it aligns up precisely to the 10 Degree arc spread of Menkaure. The value of "0" to its first level is 1.42857. Thus we find our 5.7142857 to be in total alignment.

In our chapter Inter-dimensional Physics in Volume 1 we explained that alignment occurs at a very precise 30-degree angle. As the next diagram illustrates.

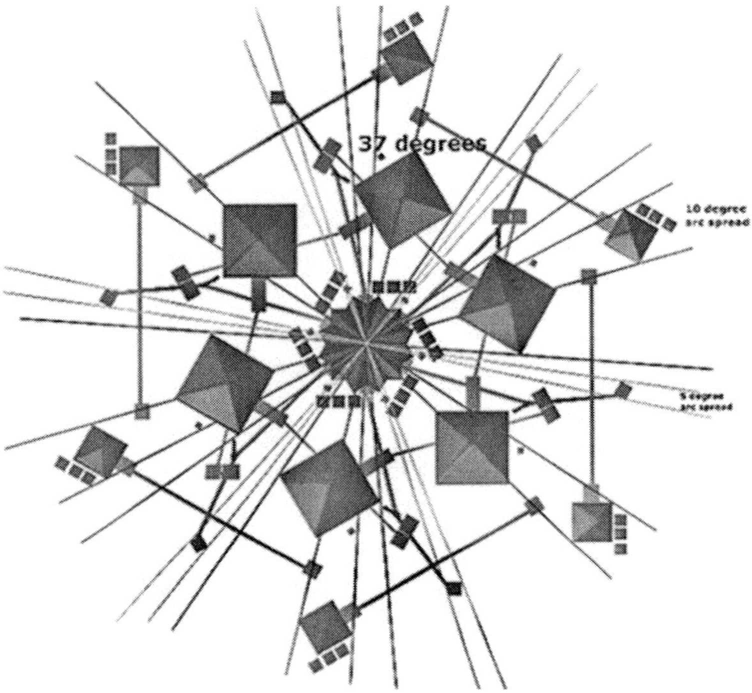

Illustration 15-5 Plateau time alignment 5

Again bear in mind that there are three lines of lineage each arriving at different 30-degree angles. Thus totalling an overall encompassing arc spread of 90 Degrees (3x30=90). Although there are four quadrants there are two halves in each quadrant, and this is defined by the presence of the justified alignments and dictated thru in the 3 Dimensional models. Hopefully the above picture will help in digesting the complexity of the model (shown in 2D). Here we have drawn back, to scale, our model and super imposed it to its maximum outside influenced point of the original model. Although you do not see it with all of the 30-degree angles, you will see the effect it would demonstrate in each of the four quadrants.

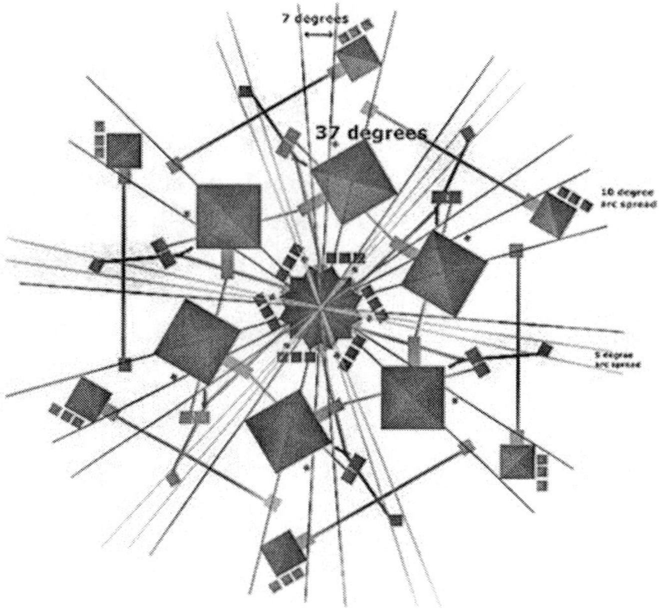

Illustration 15-6 Plateau time alignment 6

All of the above we have now shown you may help to explain how Time as we understand it must from its source come full circle in order to manifest itself in a particular timeframe. This is the reason we have witnessed the events (according to their time frame occur, but which effects the will not be felt here in this timeframe (Earths) for one full year from the date of occurrence in the Suns' timeframe.... or (10x4=<u>40</u> Sun years or the time it takes Earth to move its four quadrants through full circle)

One of the many questions we have been asked is what is the purpose of the Queens Pyramids located (three) beside the pyramid Khufu, and the (three) Queens Pyramids situated just South of Menkaure mark. When viewed in the overall model they are used to prove out the measurements (justified) of the Earths placement. In the ancient text known as Genesis we will find these measurements as those referred to as Noah's Ark. Length 300, Width 50, and set to a maximum height of 30. We may also add that there are six points with equal arc spreads of 50 degrees (6x50=300) between the two sets. As there is a 50-degree spread within the internal set, which define its max. Width.

All six of the outside 50-degree sets are converging from various 30-degree angles.

Below are three pictures showing the relevant pyramid alignments as they correspond with each segment mentioned above. They are shown as they are calculated from each of the 3 Lineages. The centre Pyramid in each lineage is the focal point of each alignment. It is also drawn to the centre each of the respective end points located on either side of the centre Pyramid. Recall if you will in several past articles, that the measurements from the centre to the outside of each Pyramid was reflected on the Plateau by the casing stones etc., as this allows for that "0" Degree buffer.

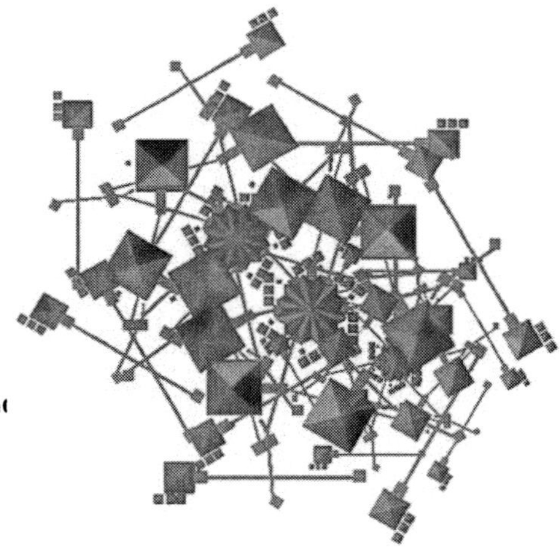

Illustration 15-7
Plateau time alignm

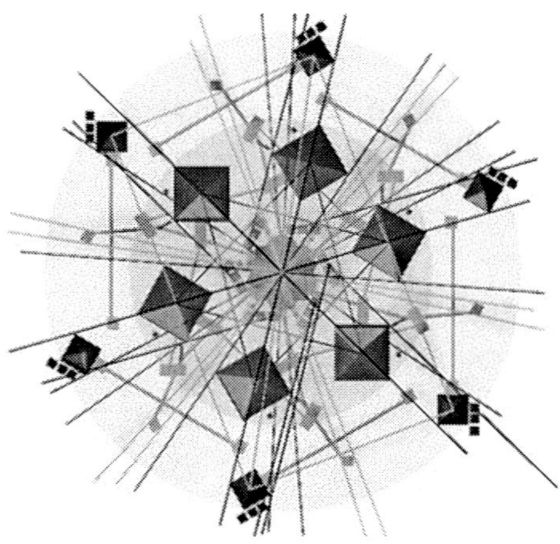

Illustration 15-8
Plateau time alignm

Another important question that is asked is what does it all mean and why is it happening now? As people have the right to know this information, and it is important to understand it, we will try to give a partial explanation to everyone. Once one understands the situation, and what created the problem in the first place, they will then understand not only is there nothing to fear, but also why it has taken until now that this rescue mission would not have been possible.

Many people have studied the various ancient texts and related information, left by our ancestors, and in doing so, has kept this data alive, but without the understanding of how all this fits together, as the material once deciphered is by itself fragmented. In other words the pieces that make the whole puzzle surrounding our past to come together have eluded us, thus this great illusion - confusion we live in today. It is this very illusion- confusion that has allowed this Dark side to keep us in servitude. And this is unacceptable. So to be able to understand that which is about to happen we must understand what took place EONS ago.

When the Earth was first populated, tens of thousands of years ago, it was set up as a FREE WORLD where all were welcome to come and go, as one would feel the need to do so. Some races even formed settlements here. Among those visitors were a group, which we will call for now the Anunnaki and to say that they prefer the Dark side would be an understatement.

This race, the Anunnaki (those who from the heaven to earth came) was referred to by Sitchin [2] when deciphering the ancient Sumerian clay tablets which were discovered buried. They came to planet earth seeking power and a location that they could settle to. In an attempt to merge their own DNA with other races, they carried out many sexual improprieties. The idea was to create a hybrid within the human race and reportedly place in a position of ruling power.[33] The Reptilian Connection, David Icke. Quote from Davis Icke website:

> There, all over the planet, you find the ancient legends and accounts of "gods" from another world who interbred with humanity to create a hybrid network of bloodlines. The Old Testament, for example, talks about the "Sons of God" who interbred with the daughters of men to create the hybrid race,

the Nefilim. Before it was translated into English, that passage read "the sons of the gods", plural. But the Bible accounts are only one of so many that describe the same theme.

The Sumerian clay tablets, found in what we now call Iraq in the middle of the 19th century, tell a similar story. It is estimated they were buried around 2,000 BC, but the stories they tell go back long before that. The tablets talk of a race of "gods" from another world who brought advanced knowledge to the planet and interbred with humans to create hybrid bloodlines. These "gods" are called in the tablets, the "Anunnaki", which apparently translates as "those who from heaven to earth came."

The ancient accounts tell us that these hybrid bloodlines, the fusion of the genes of selected humans with those of the "gods", were put into the positions of ruling royal power, especially in the ancient Near and Middle East, in advanced cultures like Sumer, Babylon, and Egypt.

Keep in mind that the crystal grid, which was laid down at a much earlier time, was working and stabilizing the environment etc. A process now nearing completion on the planet Venus.... a process you will know as Terra-forming was begun there a long time ago.[44] Venus terra-forming.

I have occasionally been known to follow my own advice. I have said as much previously in these pages that as much evidence that comes from the Martian images, someone really ought to look at Venus, and why not? If we're using the works of the Inklings as our support, no matter how unrealistic the whole contents of C.S. Lewis' Perelandra trilogy may be, there are those gems we cannot overlook, and there is both the fact that Venus, which is called "Perelandra" in those works, is the very *namesake* of the trilogy, and then there's the Templar's preoccupation with that grandest of pentagrams, the one formed by the orbit of Venus every eight years

http://www.geocities.com/Athens/Olympus/6581/double_rainbow9.html

And it is there on Venus that those who choose to remain in their human form will have the option of being placed.

But the Dark side had other plans for the planet Earth, and to make a long story short, they acquired some technical assistance from another group, which we believe to have been from the Pleiadean star system (This was but a rogue group of Pleiadians, as there are many groups within that star system). They proceeded to establish a secondary grid (we refer to this as the Red Grid) to encompass the Earth; this Grid is one existing inside of the natural more encompassing Blue Grid. This Red Grid utilizes upper Dimensional Physics. And when activated literally captured all that were within in its inner field...They were given NO CHOICE.

Now some would ask WHY with all their advanced technology they did not simply escape. The basic answer is that when the Grid was activated it came under the influence of a secondary timeline. The only way out was to either remove the Red Grid, or wait for the Timeline's to once again cross.

Well, the timelines are about to cross. And there are several (alien worlds) that are to say the least put out wanting the release of their people.

Many will recall the phrase "Let my People Go" well throughout time this request has been ignored or rejected. Soon they will come and "TAKE" their people home.

We could further explain the co-existence of these various races within and on this planet, however this is dealt with in detail in Volume 3 of the series. But the closer we get to the time, the more observable this will become.

We have all heard the expression, "my, time flies by these days" and it is well know from the latest scientific research that time is actually speeding up, or to put it another way collapsing For thousands of years the earth's pulse, or otherwise know as the Schuman's resonance, has been a constant 7.83 cycles per second. Since the early 1980's this resonance has been increasing, and is now calculated at 12 cycles per second. In "real" terms this means that there is the equivalent of approximately 16 hours in our day, instead of the previous 24 hours!

The Schuman Resonance (SR) is similar to quasi standing waves that exist in a cavity within earth's crust, and the effect of these waves seem to be related to electrical activity in the atmosphere. The properties of this cavity were originally predicted by the German Physcist W.O. Schuman in the period 1952 and 1957, being first detected (with Konig) in 1954.

In the meantime Esau is gathering forces to attempt to prevent the release of these people. By the time you read this work his forces will be busy preparing Phobos (one of the moons of Mars) to conceal his forces before it is directed towards the Earth, this scenario has been dealt with already in a previous article on inbounds (yet to arrive in this timeline). We should reiterate however that there is nothing to fear as this Phobos WILL be deflected and Earth held in its orbit. For further information please refer to the previous section on the "Dark entity".

Another way to explain this is to use the TV series SG-1 (Stargate) as an example. If you recall there is a would-be God/slave master Ra, which travels from world to world, all being worlds he controls. Equate him to Esau. Then ask yourself what keeps him on the move, in other words, what is he running from. Look at the situation this way. He is in possession of 25th level technology.

And is using this tech. To set himself up as a God to all those living, confined, to the dimensions below his level of technology. The one he runs from is the one true GOD, and that God IS the 37th Dimension. There are no dimensions above the 37th, therefore there is and can only be one GOD…and he is not HIM.

NOTES

MARS:

There has been much of late on the anomalies and "activities" reported to have been witnessed on the surface of Mars. In addition much has been mentioned of the possibility of "terra forming" in process. The reason that we are reluctant to believe that the ship parked up in the vicinity of Mars is not in the process of Terra-Forming Mars, is that the planet itself is about to evolve to its next level. This, as the picture illustrates is when the core reaches the surface (as the max. alignment of 30 Deg. is reached and when the lines cross over will cause the surface to be blown off and deposited into the asteroid belt, while the core energy is moved out into the outer solar system).

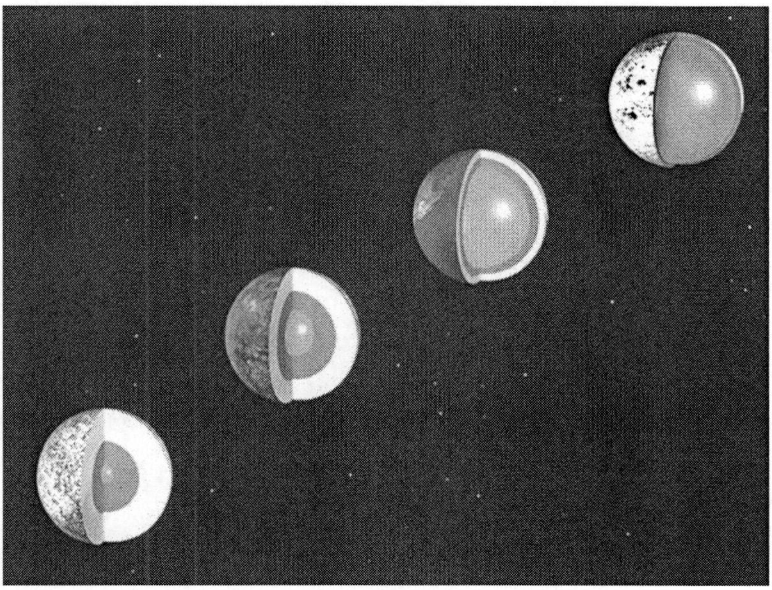

Illustration 15-9 Mars core

Illustration 15-10 (Left) Satellite image surface of Mars

http://www.msss.com/moc_gallery/m07_m12/nonmaps/M09/M0903813.gif

In this picture (satellite Image) it shows the core is getting very close to the surface and is heating up whatever underground water streams that remain to boil, which in turn creates steam build up creating a great deal of pressure. Thus compelled to exit, or relieve this pressure, by blowing its way out onto the surface. Once the steam hits the open air it will re-condense into the small water pockets we have been noticing in other satellite images of Mars. However, this picture seems to capture several aspects of this process very nicely. Therefore they would not be attempting Terra-forming on a soon to be defunct planet.

For additional information on this aspect, please refer back to chapter 4, where we deal with the Grids.

Quote
Commenting on the ACC report, Dr. Michael Wolf of NSC says

"The Russians are absolutely correct. The ETs are not just terraforming Mars, but they do have bases there, which The Cabal [rogue forces operating within the UFO Cover-Up organization] thinks premature (That's PREMATURE) to admit publically."

The "Stargate" December 28th 2000

Prior to the solar event on December 28th a massive burst of solar energy was produced which was directed in the area of Venus, thereby inducing and activating a protective field or grid around the planet Venus.

On December 22nd we witnessed our old friend the cruiser return on the C3 satellite image, while on December 25th the solar eclipse occurred.

The day immediately after December 28th when the stargate was observed, on the 29th Popocatepetl volcano in Mexico erupted, and unusual UFO sightings were observed above the volcanic activity.

[1] The Montauk Experiment, otherwise known as Project Rainbow, has been a subject of long controversy and debate. It was an attempt by the Navy to create a ship that could not be detected by magnetic mines and-or radar. In the 1930's Nikola Tesla got involved with a group that experimenting with moving through the Time/Space continuum, and at that time the University of Chicago investigated the possibility of invisibility through the use of electricity.

In 1939 this project was moved to Princeton's Institute of Advanced Studies, not that far from Philadelphia. There they were able to make small objects invisible. They presented this technology to the government and the military, because we were at war wanted to pursue it in their direction.

Tesla had come to the same conclusion as Einstein that this technology if developed would not be used for the benefit of mankind.

In 1943 the government conducted a test using domestic animals on a ship. The ship that was eventually used for the experiment, the USS Eldridge, was commissioned at the New York Navy Yard on August 27,1943 (Department of the Navy). The animals were placed in metal cages on the USS Eldridge. The ship became invisible but when it materialized many of the animals were missing or had radiation and other burn marks on them. It was thus decided at that point that humans were not to be tested.

Yet on August 12, 1943 the USS Eldridge with a full crew aboard reported reportedly underwent the Philadelphia experiment, where the men did not know what was about to happen. The generators were fired up, the switches were thrown and the ship disappeared and all seemed well.

However, others claim that the experiment took place on October 28, 1943. Substantial evidence points to the October date as being more accurate. The Navy has released the Eldridge's deck log and war diary and at no time was the Eldridge in Philadelphia. However, the records could have easily been changed.

The Eldridge's war diary reads as such: "The Eldridge remained in New York and the Long Island Sound until September 16, when it left for Bermuda. From September 18 to October 15, it underwent training and sea trials. On October 18, it left in a convoy for New York and remained there until November 1. From November 1 to the 2, it went on a convoy to Norfolk and on November 3 left in a convoy for Casablanca. The Eldridge arrived in Casablanca on November 22 and stayed there until November 29, when it left for New York again in another convoy. The Eldridge arrived in New York on December 17."

From December 17 to December 31, it travelled to Norfolk with four other ships (Department of the Navy). Although this is not the entire war log, it is the log of the ship during the suspected time the experiment took place (October 28, as mentioned above).

It would seem that the Navy never did carry out experiments on the Eldridge at any time, but the government was known to have been involved in a cover up because of national security. An example of such a situation would be the Manhattan project. This secret project was the building of the atomic bomb and no word was ever mentioned about it until it was obvious that we actually had an atomic bomb.

But the ship was gone from the harbour for about 4 hours, not just a few minutes and egend has it that the ship was transported through space and time. Four hours later it returned to its original place. There was a greenish haze on deck. Some of the sailors were on fire, some seemed insane and all were sick. Some had heart attacks, some were dead, some were part of the super structure of the ship, buried in the deck or walls of the ship, while other reports said that men just seemed to disappear and were never seen again.

But where had the ship gone for 4 hours? Witnesses placed it in Norfolk Harbour, while others say it voyaged 40 years into the future and wound up at Montauk, New York.

The Navy denied everything and said the men were lost at sea. Perhaps one day the truth will be known. The Montauk Experiment purported links several of these sailors to Montauk, New York with a time loop to 1983.

[2] Sitchin, Zecharia . One of the few scholars with the knowledge of the translations of the Sumerian tablets.

[3] The Reptilian Connection, David Icke. Quote from Davis Icke website:

There, all over the planet, you find the ancient legends and accounts of "gods" from another world who interbred with humanity to create a hybrid network of bloodlines. The Old Testament, for example,talks about the "Sons of God" who interbred with the daughters of men to create the hybrid race, the Nefilim. Before it was translated into English, that passage read "the sons of the gods", plural.But the Bible accounts are only one of so many that describe the same theme.

The Sumerian clay tablets, found in what we now call Iraq in the middle of the 19th century, tell a similar story. It is estimated they were buried around 2,000 BC, but the stories they tell go back long before that. The tablets talk of a race of "gods" from another world who brought advanced knowledge to the planet and interbred with humans to create hybrid bloodlines. These "gods" are called in the tablets, the "Anunnaki", which apparently translates as "those who from heaven to earth came."

The ancient accounts tell us that these hybrid bloodlines, the fusion of the genes of selected humans with those of the "gods", were put into the positions of ruling royal power, especially in the ancient Near and Middle East, in advanced cultures like Sumer, Babylon, and Egypt.
http://www.davidicke.com/icke/temp/reptconn.html

[4] Venus terra-forming. .
I have occasionally been known to follow my own advice. I have said as much previously in these pages that as much evidence that comes from the Martian images, someone really ought to look at Venus, and why not? If we're using the works of the Inklings as our support, no matter how unrealistic the whole contents of C.S. Lewis' Perelandra trilogy may be, there are those gems we cannot overlook, and there is both the fact that Venus, which is called "Perelandra" in those works, is the very *namesake* of the trilogy, and then there's the Templar's preoccupation with that grandest of pentagrams, the one formed by the orbit of Venus every eight years
http://www.geocities.com/Athens/Olympus/6581/double_rainbow9.html

Chapter 16
FREE WILL

In much of our work we have consistently attempted to share with our readers the understanding and results of long-term research concerning the complete Giza Plateau pyramid complex. In addition, we have demonstrated the interwoven and definitive tie in of the Books of Genesis and Exodus with the exact location and construction of the Giza Plateau. The results of which have given us an incredible insight as to our very origins in this cosmos and future events about to unfold. Results that corroborate many similar fields of research dealing with the biblical and scientific aspects of our "end times". In particular the 2012 date, also known as the "Mayan calendar end count", which we proved out utilizing the layout of the Pyramid complex at Giza showing us the date to be December 21st 2012, at 22:18 local Cairo time. In chapter 15 we ended with a summary of the breakdown on the Timeline leading up to End Time.

We have explained all these events in our previous work, however, what we have not as yet explained fully is where (or actually from who) certain information originated. Due to certain events that had to be allowed to pass in recent times, we were unable to reveal HIS real identity. In this vital update we now reveal who HE is and the purpose for HIS arrival on the scene at this most important point in earth's history.

On May the 5th 2000 a Star alignment took place, this alignment afforded the opportunity for a Time Portal to be accessed. This alignment took place between the Star known as Betelguese in the constellation of Orion (Biblical reference is Betel) and the Sun, through to Earth. At the time of this alignment several people here set about carrying out their assigned tasks, all of which were completed successfully. This work allowed for the return of the

Entity that you have read as referenced as Joseph, or sometimes affectionately as "Unc". This entity was returned to **"His"** body, which has been closely guarded (awaiting **"HIS"** return) beneath the Giza Plateau. Since **"His"** return to Earth **"He"** has made his intentions very clear to Esau; **"He"** is here to **"Take back his people". "He"** has now left the plateau.

Therefore, in light of recent events it seems timely to reveal information on the soul and what it means to make a Free Will choice. For this "Free Will Choice" can only be made when the individual knows what the Choices are, from which one must choose. It is equally important that the individual should understand that GOD empowered each Soul with the individual responsibility, and right, to make that choice.

THE SOUL AND FREE WILL CHOICE

At this point in time, travel through the grid may be carried out on a physical basis of body travel within the Earth Grid (Blue Grid). At the border of Earth Grid - Universal Grid, the human body must be released, and only the soul can travel beyond. The reasons for this can be explained in the chapter on time (chapter 15). In most cases, even travel through the Earth Grid is non-physical with the body remaining here on Earth; as there is no need for a 3 dimensional body when out amongst the grids.

As thought is used to direct travel through the grids, it becomes obvious that thought must be connected to the soul. What we think and feel as humans, continues past the human body. It therefore follows that as intent directs thought, and Free Will directs intent, these are also connected to the human soul.

Free Will implies choice. In all our previous research, we have indicated that people must now make a Free Will choice. This means we must make a soul choice.

Being humans on Earth, we really ultimately only have two choices as soul choice. There are only two ultimate sides in the battle…God or Lucifer (**Esau**). Every choice we make, therefore, eventually becomes a choice for one of these two, as the battle is between these two.

Lucifer wants Earth. He wants Earth because he wants us, God's people.

And he wants us because we, being of covenant with God, possess the power to go back to our Source (God). Ultimately, Lucifer wants God's position, and he thinks he can get that by possession of our power. He wants our thoughts, our intent; our Free Will choice....he wants our souls.

So how DO we make a Free Will choice? Choice is made by setting intent and once intent is set, thoughts flow to God, and our will becomes one with God's Will. That flow is a two-way flow, in that we then will take action from the flow of God's Will to us. That connection is the connection of our souls to God, and upholds our covenant with God.

It follows then that if we change our intent from that of God's will, we break our covenant with God, and the price is our soul.

As our soul is destined to return to our Source (God), in accordance with the covenant, breaking that covenant and changing our intent does not mean our soul will go to Lucifer. What it does mean is that our soul does not return to God. In effect, our soul will stay right where it is at the time of breaking the covenant...it does not progress, it does not complete any cycles, it does not return to God; just stays where it is. In effect, by breaking the covenant, we create our own "grid lock".

The battle of Armageddon is taking place to release this gridlock we have placed on ourselves, as well as the gridlock that Lucifer has tried to place on Earth. It will be fought between God and Lucifer because Lucifer thinks he can own our souls to return to his previous position near God. Lucifer thinks he can be God.

Through this battle God will again make us aware of our covenant with Him. He is giving us the Free Will choice to continue our path of returning to Him, or staying right where we are now.

Lucifer cannot win this battle, we already know the outcome, however, Lucifer does not know he cannot win. We have a covenant with God that Lucifer never had, a covenant brought to us by one we know as Jesus Christ manifesting into physicality...a covenant that cannot be destroyed.

But, we do have the choice. We can give our will over to Lucifer and do his bidding, but even if we do this, our soul cannot be given over to Lucifer.

However, if we do give our will to Lucifer we create our own gridlock, whereby we ourselves cannot go back to God until we change our intent. Lucifer cannot "ride on our backs" back to God. Whatever choice we do make, our soul remains connected to God.

The individual's choice will determine whether that individual soul will continue to a higher plane or stay where it is. The choice to uphold God's will enables us to go on to a new cycle, a higher dimension, ever closer to God. The choice to align with Lucifer's will means to hold ourselves right where we are now. Not making a choice will also hold us here, for this battle requires us to make a choice.

Please bear in mind that this choice to continue or not continue the path to God, is quite different to the choice required for being lifted or staying for the battle. People of the covenant with God, have already made the decision long ago to take the path back to God. They now have a choice of whether to fight the battle or not. Either choice will not negate their choice of path, as long as they have stayed true to the covenant. It is the choice of remaining true to the covenant or not, which decides whether the individual continues on to higher planes or remains in gridlock. An individual not previously upholding the covenant, who asks to be lifted, also is making a choice to now return to the Covenant and the path towards God. Those who do not have the intent of God's will may not be lifted. Those choosing to remain to do battle, who have not upheld the covenant, can make the choice to return to the Covenant at any time, but will remain until the battle is ended. Thus from either aspect, a choice is required.

The message to Lucifer "Let My People Go" means that Lucifer also has been given a choice. The red grid can in fact be taken down by Lucifer if he so wishes, alternatively God will destroy it. If Lucifer chooses to drop the red grid, he would be acting in accordance with God's will and go on to his next cycle of redemption, through his own intent. But Lucifer will not do this, as he wants to be God.

That only leaves one choice, and that is the destruction of the red grid, through an almighty battle. By holding his original intent, in effect, Lucifer is giving us a choice, and our choice can bring us closer to God.

In our section **A Given Sign** we explained that we had asked if **"He"** would provide **"His"** people with a sign. **"He"** was initially reluctant to do this (only now do we understand the reasons for **"His"** hesitation to do this at that time). **"He"** did however eventually work out how **"He"** could do this without causing a deviation to **"His"** overall plan. "**He"** then instructed us to watch the satellite images and that **"He"** would provide a suitable sign. This sign was not intended to be visible from the surface of the planet.

Imminently (the exact date or circumstances unknown) the person whom we have been in contact with for some time, the one we have called JOSEPH, **has informed us that we will witness an** "apparition"**. Some will see a crucifix (Crucis) while others will see a Star of David.**

We have previously called this person JOSEPH, under instruction from HIM to do so until such time as we were requested to reveal His **name. Although we have known HIS name and identity, we were not at liberty to reveal it. HE has now given permission for us to reveal that** His **name is** ... Jesus Christ**. In** "**His**" **words** "Let it be known these are My sign." **When asked when we should reveal this information,** "**HE**" **replied** "**NOW**!".

We are well aware of project Blue Beam, this will NOT be a holographic image, and we believe that the people who are aware will understand and recognize the true sign.

This will be a sign for the call to arms for "HIS" people, and many will awaken at this time.

He**, with many helpers is preparing for many major earth events (earthquakes and other similar events) where many souls will be leaving earth requiring guidance and council.** HE **and His helpers will be watching for some time, and when certain events come about will allow a few major players to enter the arena.**

We must make this very clear:

Once the sign has been given many, many souls will be awakened to this call. For this IS **their call to arms. To each entity this will have various meanings, as each has his/her own task to complete. Others have approached "Him" and requested that the length of time the "Trials" last shall be shortened.** "His" **response was that it totally**

depended upon the response to his call.

Preparation

When we asked how to prepare for the coming times, "His" **response was as follows.**

"Be steadfast in mind and soul, prepare as best you can for the body, have all your spiritual tools ready, and be ready to heal others."

When asked how this information will be received by others **"His"** response was as follows:

"Not all will understand, even some of my faithful will not understand immediately. In hindsight they will understand though. Many will awaken. And those who laugh now will remember later."

Stargate

Our calculations and research of a "stargate" opening on June 10th 2000 were confirmed, and we have been told more will enter than leave. However, this does not necessarily imply a vortex in our protective grid, and events will be determined on other events in progress.

In addition, much will depend on the answer to His **call.**

In confirmation of the above, we refer the reader to the previous chapter (15) dealing with "time" where we explained the phenomena of the one year delayed reaction of the witnessing of events and the result of the effects itself. In addition, in chapter 8 we explained the alignment that occurred on June 10th 2001 (being one year exactly from the event of June 2000), thus demonstrating the effects of timelines where one timeline crosses over with another. Each time a timeline crosses a chain of events occur in that timeline and those events, until they have occurred in that timeline, cannot effect the subsequent timeline until they have occurred themselves.

To demonstrate this point The Book of Revelation refers to the 1000- year reign, which we explained earlier in the book as 10, 100 and

1000, where the outer timeline is represented as 1000. When this outer timeline comes into alignment with the middle time line (represented as 100) and events are realized, then the events in the inner timeline (represented as 10) come into existence and the cycle is complete.

We are of the opinion that this is the very reason that we are warned against attempting to decipher the book of Revelation itself, which is written in a style of the worst-case scenario. However, we also believe that the trials and tribulations we will undoubtedly experience will be dependent on the combinations of actions and perceptions of humanity, thus in turn guiding our destiny.

Due to the timeline process, the actions that occur in the timeline prior to our existing timeline would have to be prejudged based on the best possible information available at that time. In other words, certain criteria will have to be met on the available information that will initiate an action. The resulting effects of that action we experience in the form of reaction may often be referred to as the "trials and tribulations". This is the very reason that we feel it is imperative that humanity must imminently make their choice known, in order that the more people that have chosen correctly will determine the lengths and the depths of the tribulations (actions) that humankind will undergo. This will assist us in our attempt towards ascending to the next level, the alternative being that we may not be allowed to make that ascension forward, thus this worst case scenario would in turn bring into play the worst case action.

Chapter 17
THE MAGNETIC POLE SHIFT

Shortly, there will be a shift of Earths Magnetic Energy field this in turn will advance our Time Line. Several changes will occur as a direct result of this shift in energy, particularly occurring on the earth's outer crust due to the alterations as our planet makes its necessary adjustments down, and through towards the core. This sharp shift in the adjustment will cause our polarity (pole axis) to shift very quickly. Many events will come about during this **controlled** shift, or as a direct result of this shift.

Thus, in this section we explain how this shift will affect the Spirit or Soul when this occurs. We will cover such topics as the weighing of the Heart, the reading of the Book, and explain exactly what the circumcision is, and how this ties directly into our personal responsibility in carrying out our part of our Covenant with **GOD**.

As stated above this will be a **controlled shift**, we will now explain why this is a necessary event and **MUST** occur, within a very narrow window of opportunity.

ESAU would see this world and it's inhabitants completely destroyed, rather than allow us the opportunity to exercise our **Free Will** choice, that being to move forward following **GOD's WILL**. God's will, therefore his Law, allows for the physical in addition to the non-physical to co-exist in a harmonious fashion. The heartbeat of the physical Earth itself is presently out of sink with the Galactic and Universal heartbeat. This occurred as a direct result when the Dark side encased the planet within the artificial "RED" Grid. What this Red Grid did in effect was remove this planet so that it's heartbeat was no longer allowed to grow and be strengthened, or keep pace with the rest of the Universal bodies, of which it was previously beating in harmony with. This soon to be realised 'controlled shift" will be executed in order for the Earth to swing back into the Universal harmonic pulse timeline.

"IF" this controlled shift were not to take place prior to Dec. 21st 2001, the Earth, as we know it, would be destroyed as it approaches the present "Universal" timeline, which it must do in order to prepare itself for the transition, which it must undergo prior to the "End Time" of Dec.21st, 2012. This is the date that the "Universal Heartbeat" will once again begin, requiring all that must be in harmony at that time. In Earth time the, "0" time is the time relative to Earth which will pass is equal to the pause between the present Universal heartbeat and it's next heartbeat. Keep in mind that it is a full year (one complete rotation) both going into this "0" time as well as exiting it.

At the present time the Earth is under extreme pressure as a result of the very existence of the "Red Grid", having removed the planet from it's natural ability to adjust itself under normal growth rate etc. When the "controlled" shift takes place, the entire world will shake and vibrate violently. In addition we will witness that TIME itself will seem to come to a standstill, this aspect will be short lived however, as the Earth swings back into the Universal Timeline. If this shift were left to occur on it's own in December, this shift would definitely cause the planet to dislodge itself from it's present orbital path. However, as previously mentioned this event will be a very "controlled" shift, and much help has been brought into the inside of the "Blue Grid" in readiness to stabilise the Earth, and allowing to maintain its orbit. This help is presently sitting in readiness in the 4th and 5th dimensions, making them impervious to any attack from the dark side working within this 3 Dimensional world.

This particular assistance entered into our space immediately prior to the large quake that occurred in India at the end of January 2000. As these people were "Lifted" and rescued, it was found that their SOULS were extremely distressed and required much counselling. It was then decided that people must be made fully aware of what is happening so as to help them by removing this FEAR through understanding. On the light side there is no fear present as all is understood and the Dark side is forbidden to enter. It is imperative that those whom have made their free will choice emit a feeling of Love rather than Fear, for those emitting Love will be easily recognised and lifted quicker (bear in mind that when you get that disoriented feeling just relax – let them do the work). Those who emit fear make it much harder for those helping with the lift to know of your choice and consequently they must then read the AURA, to determine one's intent.

The only other point that may hold one back is Negative Karma. This negative karma is caused by the failure to have dealt with past events which one's soul is therefore still connected to those events. To dispose of this Negative Karma it is as easy as asking the following: Forgive us our trespasses, as we forgive those whom have trespassed against us.

Over a period of several years many people have been given implants, these implants were embedded into various people for the purpose of control, an alien technology compliments of the Greys'. A negative group of humans has used the Montauk program to facilitate the procedure by allowing them to remove the subject into another time frame, implanting them, then returning the subject back into the subjects' real time (but only a few seconds later). This is not so hard to believe when one questions events such as HOW do they get to a downed crash site seconds after a crash, in addition no one actually sees them coming, or going!!

These implants are generally implanted for one or two reasons...the first to control all those who work unknowingly on these Black Projects. This also includes those in the public sector who would have come forward and made public that which THEY did NOT want anyone to be aware of. This same "time hopping" has been implemented to divert public attention from any news releases that were destined to draw public awareness to the events going on around us.

Most of the Greys have vacated this planet, as we explained in a previous chapter in volume 1, there is however, a small contingent remaining presently housed in Nevada. They will be dealt with soon enough. But one of the advantages of this Energy Shift will be that all the implants that they have been placing into peoples' bodies, will no longer be functional. MANY will AWAKE!!

This Energy Shift has already occurred and was successful! However, we have not arrived into that Time Frame yet, but it occurred during the time that ESAU began to dig into Phobos, and to help to explain the situation an interesting and extremely relative article may be found under the title; THE STRANGE CASE OF FOBOS-2 ("Fobos" being the Russian spelling)[1]

By Jim Oberg
Special to SPACE.com
posted: 05:50 pm ET
30 June 2000

This past decade has not been kind to Earth's Mars probes. There was NASA's expensive Mars Observer blowing up in 1993 as it warmed its rocket engines up to slow into orbit. And we've just seen both of NASA's 1999 missions fail.

Russia lost another ambitious probe in 1995 when its upper stage failed, dumping radioactive fragments onto the Andes Mountains. And a Japanese mission, their first to Mars, went off course right out of the gate in December 1998.

But the most bizarre loss of a Mars probe is unarguably the case of Phobos-2 (or Fobos-2, in the Russian spelling). It "disappeared" in March 1989, under very unusual circumstances that still mystify and excite many people.

Recent developments in the Russian space program have opened new insights into that failure. But first, here's some history.

Trying to lift the curse

The Soviet Union launched two probes towards Mars in mid-1988, trying to break a decades-old jinx. Its initial series of small probes (1960-1965) had been a total disaster, and a series of heavier probes (1969-1973) didn't do much better. But this third generation was much more promising.

The spacecraft "bus" — the main body — was of an entirely new design. It had new engines, new computers, new communications gear. And this new mission carried sub-satellites to be dropped onto the inner Martian moon, Phobos.

But the old jinx still prevailed. The first probe was lost due to an erroneous command on the outbound leg. The second vehicle was crippled by electronics failures and by the time it reached Mars on January 30, 1989, it was operating on its last and lowest-powered radio.

Nonetheless, it slipped into orbit around Mars and slowly matched its path with Phobos. As it closed in, it also made observations of Mars.

A dozen times, it turned its cameras away from Mars and towards Phobos. This required the whole spacecraft to turn, since a movable "scan platform" hadn't been installed. The manoeuvre also turned the dish antenna away from Earth, cutting communications for several hours each time.

On March 27, 1989, the probe began another Phobos photo manoeuvre, and as expected radio signals ceased. But after the planned manoeuvre, when listeners on Earth expected to reacquire the signal, nothing was heard. More careful listening picked up brief bursts of radio signals, as if the dish antenna were swinging wildly through space and only occasionally beaming back towards Earth. Then - only silence.

Strange shadows explained

But not for long. Soon a strange and wonderful story grew and spread, about mysterious structures observed on the surface of Mars. The probe's last view, so the tale continued, showed a miles-long oval object closing in. The object's elliptical shadow could be seen on the surface of Mars thousands of miles below.

And indeed, cigar-shaped shadows were plainly visible in many of the 37 photographs that the doomed probe sent back to Earth during the 60 days it survived circling Mars.

Such images are not unusual in the archives of American Mars orbiters, from Mariner-9 to the Viking Orbiter twins, to the Mars Global Surveyor, still at work there to this day. In those cases, what was seen were shadows of the moonlet Phobos, stretched by being projected at a low angle to the Martian surface.

But Fobos-2 was on a different orbit, and for the last few weeks it was fairly close to the moon Phobos. Thus, any view of that object's shadow on Mars would have to be fairly circular. Think about the geometry — use props if you want to. It's an illuminating exercise in 3-D perspectives. If you're next to a roundish object casting a shadow, the shadow will look more-or-less roundish to you no matter how steeply tilted the surface it's projected on is lying.

So where did the cigar-shaped shadow come from?

Several years ago, mission scientist Aleksandr Selivanov explained the cigar-shaped shadow in the Fobos-2 images this way.

The imaging system is a "scanning radiometer," not a camera, Selivanov pointed out. A rotating mirror moves perpendicular to the line of the probe's motion over Mars. As a result, "a picture is generated by the motion of the spacecraft in its orbit." The probe did NOT gather an entire image in one snap, but accumulated it over a period of time, line by line.

Fobos-2 was staring straight back along the Sun-to-Mars line, to get the best infrared readings. In contrast, visible light imagers prefer to look for shadows cast by surface features, so they are aligned at large angles to the Sun's rays. This made the visible-light images from Fobos-2 look washed out.

Now, Fobos-2 was quite near the moon Phobos in the last days of its flight, both circling Mars along the same path. So the roundish shadow of Phobos was on Mars's surface, within the field of view of the scanner, when the scanner was looking "down sun" at Mars.

Selivanov explained that if the probe had been rock steady, the Phobos shadow would have left a dark streak right through the entire center of each image, as the image was assembled line-by-line over the course of each orbit. Because of a slight rocking of the probe, however, the scanning beam "sliced" the Phobos shadow at different points, from back to front, over the course of each imaging session.

The resulting elongated shadow is thus an artifact of the imaging technology, and of the probe's motion through space and around its own axis. Selivanov argued that since these shadows are all precisely aligned along the probe's flight path over Mars, they are unquestionably not shadows of other objects near Mars. They show the shadow of Phobos.

One supposed photograph, the "last one before the attack," shows Phobos and a bright vertical line below it. Since the line runs right along the telemetry scan lines, space experts are confident it is some sort of transmission flaw, not a real object in space. Besides, the date on the image is March 25, two days before the probe's loss.

Hurry and other human errors

The breakdown of Fobos-2 was disappointing to experts associated with the program, but not surprising. They had seen human error doom its sister craft, Fobos-1, before it even got to Mars, and they had seen signs that Fobos-2 wasn't in much better shape.

Dr. Larry Soderblom was one of the American scientists who had instruments aboard the probe. "There is a feeling in the American space science community the Russians were in too much of a hurry," he later told a reporter. "The two satellites lost were launched without much thought to a system of checks and balances that might have prevented such problems."

And even the design was questionable. "Soviet scientists at the Space Research Institute in Moscow complained that the new, sophisticated spacecraft actually was designed for purposes other than those for which it was being used on the missions to Phobos," wrote a British space expert. "Engineers adapted it for the mission in order to flight-test it for future missions to which it was considered better suited."

Here on the planet itself the Dark side pawns (The NewWorld Order or NWO) are instituting a collapse of the monetary market, and as history will tell you this will help them to launch a WAR. This scenario, as they will soon find out will not go as they planned.

A few of the 2 Dimensional diagrams we will be utilising in this chapter are borrowed from our work on the DNA, and how its process is calculated (or spelt out) once the overall Giza model is deciphered. Although we would suggest that readers work with the diagrams to help visualise the material, we caution that some of the illustrations may not be those best suited for exactness in relation to the material. But these were stock material pictures that sufficiently offer the visual as it relates to the topics discussed. The variance comes about due to the capture of the angles of the 3Dimensional model. We have attempted to use only the pictures which were top views, as they offer the best angle for 2D viewing, however some of the angles are shot from offset angles. We might also reiterate that the measurements and placement of all Pyramids, Temples, Causeways, etc. located on the Plateau, DEFINITELY support the information, which we now put forward.

Pi and Phi

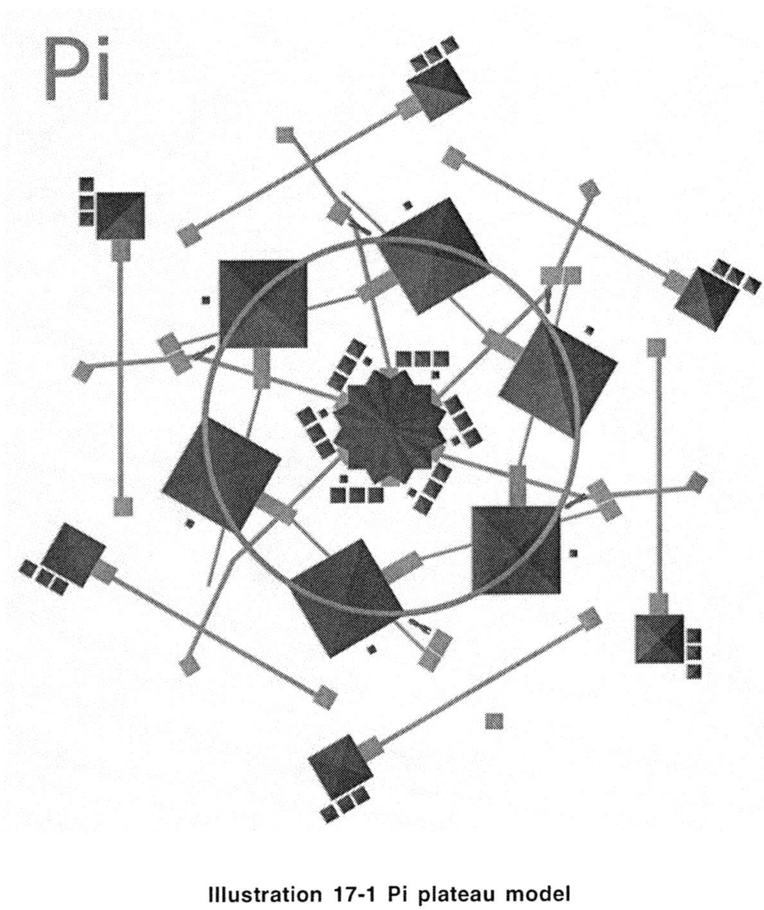

Illustration 17-1 Pi plateau model

Both Pi as well as Phi is precisely depicted on the plateau using the FULL model, in that the precise size and location of the various structures etc. mark VERY SPECIFIC points.

Khufu- centre point (point in question)

Khefre- represents the pi element as it relates to Khufu.

Menkaure- represents the phi element as it relates to Khufu.

The Queens Pyramids- of both Khufu and Menkaure are used to show how both pi and phi are incorporated into the Duality of the Living Earth (and/or man).

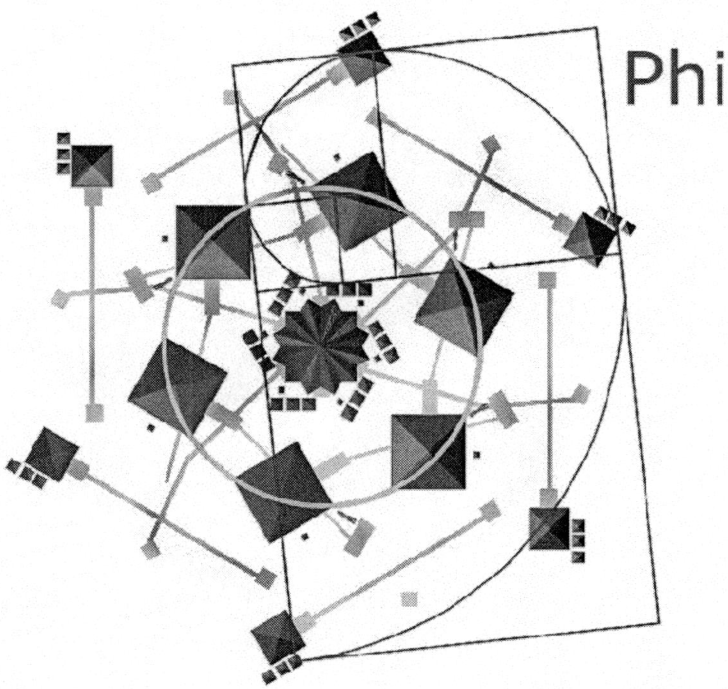

Ilustration 17-2 Phi plateau model

DUALITY

The Sphinx that sits just above the PI is a fair representation of MAN in his Duality. The depiction should be viewed as follows; The Body of the Sphinx is that half representing the animal/beast in ourselves, and rising above the physical body of the beast is Man. Meaning that the physical (or the Beast) is the equivalent of PI, in that it's cycle is that which is born –lives and then dies. The Head which rises above the "Beast" is that representing the SOUL of MAN, that which only the PHI can breath the Breath of Life into the physical body of the Beast. Pi being a closed system and PHI a forward spiralling system beyond that yet incorporated into the PI cycle. This is not to say that animals are without souls or life force, for they are all Gods creations. Whenever the Double Helix is present in the DNA makeup the Duality is present.

Many years ago we were informed through one of Edgar Cayce's readings, that the Hall of Records was located BENEATH the RIGHT paw of the Sphinx. This is very important as well as being VERY symbolic, that being the history of man laid beneath the Right paw verses the Left. There is a HUGE distinction to be drawn between the left verses the right.

The distinction is contained within both the Biblical texts and Egyptian symbolism – as depicted through hieroglyphic and the diverse visual Egyptian art forms.

As a demonstration relevant to our article, the prime example used in the Book of Genesis is non other that the story of the birthing of ESAU and JACOB.

> Genesis 25:25
> Now the first came forth red all over like a hairy garment, and they named him ESAU.
>
> 25:26
> Afterward his brother came forth with his hand holding on to Esau's heel, so his name was called Jacob; and Isaac was sixty years old when she gave birth to them

As the Beast was born first, that being Esau, clutching onto his right hoof /foot clung Jacob as he too was pulled from the womb. This is the story of the History of the Duality of Man, first the PI (or Beast) emerges and due to very existence of Pi it is there that PHI comes into being…thus the duality of man begins.

The Right side represents the "BEAST" for it is the physical flesh that is first created. The Left side follows and with it carries the other half of the Duality, that being the "Breath of life" or the "SOUL. Thus it is found that the History of man would lay beneath the "Right" foot, in this case the right paw of the Sphinx. The next is the Left, to symbolise the potential future consciousness of man, as the Left foot is placed forward on its journey towards Ascension.

One will notice that in most Egyptian statues of standing figures they are depicted with the "Left" foot forward. This symbolism is to help the subject "GET OFF ON THE RIGHT FOOT"- that foot is the LEFT foot towards its journey in the afterlife. This leaves our animal or physical body behind on one's journey to Heliopolis (City of Light).

Later in this chapter, we will be drawing our attention to the Egyptian symbolism as it is found in Egyptian art forms.

Illustration 17-3

Menkaure with the goddess Hathor on his right and the personification of the 17th Nome of Upper Egypt on his left. It was discovered in the valley temple of Menkaure at GIZA in 1908.

SACRED GEOMETRY AND BIOLOGICAL IMMORTALITY

A highly relative piece of work to our ongoing research at Giza is the above title by Dan Winter.[2] Dan Winter explains how sacred geometry and biological immortality are connected by PHI, or the golden mean ratio. It has been discovered that PHI has been found in the frequency between harmonics of the heart at the moment of sending love. In addition, it is found in the frequency between the harmonics of the heart at the moment in the ratio of the moment of maximum pressure in the shape of the hug or squeeze you give to send love.

To make a succinct overview as related to our overall related research, the PHI wave is and can be embedded on other waves, or lineages. For example the earth wave (lineage) is embedded on our sun's wave (lineage), which in turn is embedded on its star point of origin and so on. The result is recurrence and perfect sustainability; therefore, the key to our ascension process is the perfect wave of love – phi frequency –, which brings us to the ancient Egyptian ceremony of the "weighing of the heart".

In this ceremony, as we have previously discussed in detail, the heart weighs lighter than the feather and thus the soul of the deceased is "allowed" or "qualified" to pass through into the Duat, the Afterworld, where Osiris reigns. This then is the embedding of our souls, we embed our souls to the creator and in so doing one carries the frequency of our ancestors within our own soul.

Thus in effect is the process of how the covenant, and our lineage are fully carried forward onto the next level. This "chain" can never be broken, but on occasions may be halted at any point in time/space by the mere decision of "soul choice".

Keep in mind that the human heart is located on the LEFT side.

In a previous chapter, **Lift up Your Hearts**, we made the stunning revelation of the true identity of the entity who has been our source and guide of much information since HIS awakening under the Giza Plateau on May 5th

2000. We also pointed out that a given sign will be witnessed and discussed the understanding of the soul and free will choice, and so it is now in this update where we will offer an explanation of the process that will set our soul energy free.

> Thoth, the judge of right and truth of the Great Company of the Gods who are in the presence of Osiris, saith: Hear ye this judgment. The heart of Osiris hath in very truth been weighed, and his Heart-soul hath borne testimony on his behalf; his heart hath been found right by the trial in the Great Balance. There hath not been found any wickedness in him; he hath not wasted the offerings which have been made in the temples; he hath not committed any evil act; and he hath not set his mouth in motion with words of evil whilst he was upon earth.

THE SPEECH OF THOTH:

From the BOOK OF THE DEAD…or PAPYRI OF ANI

In one of our previous chapters in volume 1 THE EIGHTH DAY we also included the following;

Genesis

17:9

> And God said to Abraham, "As for you, you shall keep my covenant, you and your descendants after you throughout their generations.

17:10

> This is my covenant, which you shall keep, between me and you and your descendants after you: Every male among you shall be circumcised.

We explained at length the meaning of the above and its connection with our lineage, where the circumcision as relates to Genesis, symbolises the covenant made between Yahweh and his people. This symbol is often found in much 'older; civilisations, but in Biblical Hebrew it literally means to cut a covenant.

We also gave this message from one Euphraim (Joseph's second born)

THIS MESSAGE FROM EPHRIAM....

> Send my message to all those who now suffer. I am here to bring FREEDOM. FREEDOM and FREE WILL are one. A message also to those Tyrants, "you know who you are" you will be suppressed as you have suppressed others. Be warned...the time is now. The choice is yours. The choice is FREEDOM.........

And from OSIRIS...

> We love you...regardless of everything else we will push through the Love

THE COVENANT
THE BALANCING OF THE HEART AND SOUL

Over recent months many have been enquiring as to the purpose of and the "qualification" for being of the COVENANT. In this section we will attempt to offer a brief overview in order that one may at least grasp some modicum of understanding, bearing in mind that this is an extremely complex scenario, where we do not profess to have all the answers.

Over the aeons of time our true ancestral 'origins', due in part to the many genetic interventions and variations, may never be fully realised. However, by reading through the ancient Egyptian and Biblical texts, and applying the analysis that we consistently followed, we may be able to open a crevice towards an understanding of this subject.

Before we submit our understanding of what people term the COVENANT, a few pointers may be helpful.

We must initially point out that we do not really know who's names are written in the book of the Covenant, it is basically a personal thought on belief or faith. However, it is believed that in the main it would be the original tribes of what we call 'human kind', those possessing human DNA. In addition, those people who agreed to keep the spiritual connection, and have continually done so, including those who have asked to be saved and forgiven.

Which brings us onto our last point, those whose hearts balance out to the feather of truth, Mayet.

The meaning of the weighing of the feather:

Many will recall the Testament advice that there shall be NO OTHER GODS BEFORE ME. NO CRAVEN IMAGES and so on. This is because MANY have forgotten the Covenant with HIM, they still have yet to believe in the ONE GOD!! MANY follow or worship false God's and Deities, and yes, even Money is worshipped. Thus, this worshipping of these false God's adds weight to the HEART...Until one truly lets go of these false God's their individual Hearts will weigh more than that of the feather....

Once these initial events have taken place, the "SEVENTH SEAL" will be opened and all those wishing to depart this time event will leave. Others will then arrive and the seal closed.

The seventh seal will be opened no later than December 21st. 2001 as this is the point in time that the window of opportunity to do so will close. This being when the Sun's "0" alignment and the Earth's "0" alignment time will intersect. The following Diagrams are in addition to those found in our chapter 15 dealing with time, but in addition are from our work on the DNA modelling.

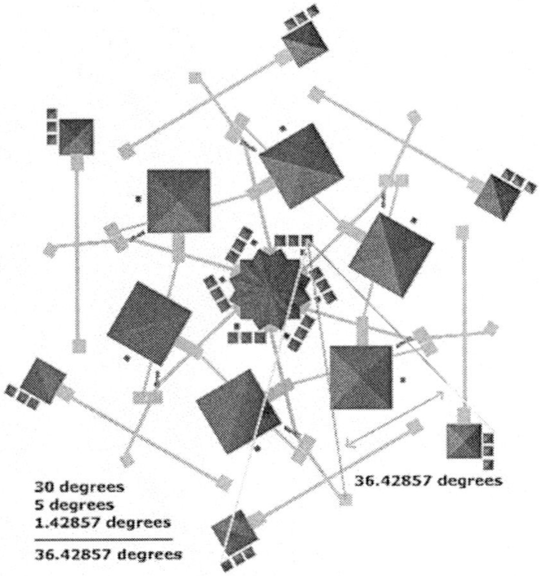

Illustration 17-4 Plateau model DNA1

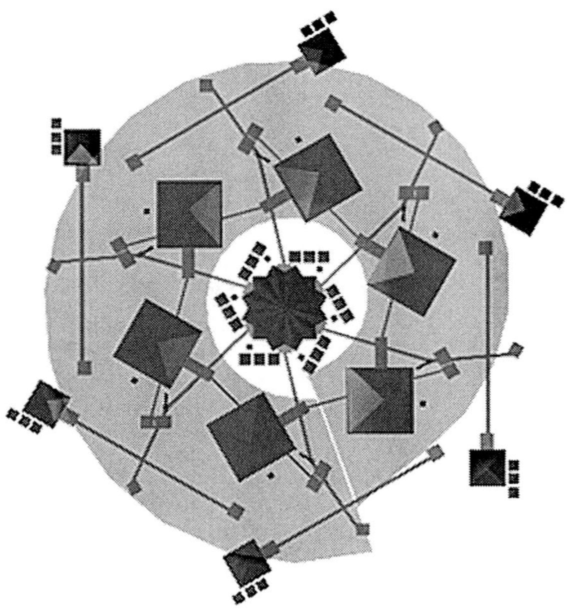

Illustration 17-5 Plateau model DNA2

Illustration 17-6 Plateau model DNA3

Illustration 17-7 Plateau model DNA4

An interesting source of information concerning the comet Kahoutek comes from a letter sent to NASA from Marshal Masters of the Millenium science group in July 2000, the letter is included below

NASA Responds to Our
Open Letter Regarding
76P/West-Kohoutek-Ikemura
June 23, 2000

TO: Charles S. Morris of NASA
FR: Marshall Masters

Dear Mr. Morris

Thank you for the reply in which you raise a very significant point: "There has been no ephemeris posted on this comet because it is not currently observable."

While you did provide an updated ephemeris, is obviously based on a sightings of 76P made prior to the recent 76P flyby of Mars. The numbers therefore are based on assumptions as opposed to observed facts which is understandable. As you point out in your reply, "Hubble can not observe it either because it is too close to the Sun."

Once Mars, it's moons and 76P can be observed, then and only then can anyone claim with any certainty that this "much to do about nothing." As to the other issues raised in your reply, the readers can judge for themselves.

Marshall Masters
The Millennium Group

For another view on this subject refer to the Australian Priest know as "Little Pebble" [1]

THE FLAIL AND CROOK - SLAVERY OR FREEDOM?
(The Mark of the Beast)

Here, we will explain the relevance and importance of the "Royal Regalia" (the flail and crook) often seen in figurines of Osiris and that of the representation of Pharaohs on their outer coffers, where they are seen to hold these symbols in either hand. It is the method and by which hand they hold and represent these symbols that we will discuss below.

Illustration 17-8 Osiris Figure

The main concept that we intend to explore (and for you to understand) is the idea that the flail and crook (traditionally considered to be symbols of government and authority) are in fact symbols of slavery and freedom, opposing concepts of the spiritual. We then go one further and suggest that the flail, a symbol of slavery, may have its origin in what the Book of Revelation termed "the mark of the beast" or the number of man.

However, to understand the complexities of the depiction of art in ancient Egypt, and therefore to come to an understanding of our conception one must understand the contemporary concepts of daily life, afterlife and culture.

FRONTALITY IN EGYPTIAN ART

Illustration 17-9 Osiris Statue

It must be realised that the style and purpose of Egyptian art make little sense without a detailed understanding of the culture. Essentially, all art forms were functional, whether painting or sculpture, and were concerned primarily with the continuance of life. That is, not to reflect reality but to replace and perpetuate it.

The aesthetic achievement varied, but the purpose remained the same – to represent, influence and manipulate the real world.

The portrayal of each individual element of a subject was made from the most representative angle. For example; the human torso and the eye were best viewed from the front, while the arms, legs and face best seen from the side, thus confusing the overall effect. This often causes Egyptian depictions of human figures to appear distorted and inconsistent to the modern eye. This is also evident when we come to view depictions of figures in relief on stone and Pharaohs on outer coffins, where we observe their arms in a cross over position covering their chest. In this pose the flail and crook is also evident, being held in each hand, and it is this representation of which hand each is held brings confusion to many.

Like Classical Greek statuary why then did they not discover the geometric rule of perspective, and why was left and right muddled? This has nothing to do with the skill and imagination of the artist, but everything to do with the purposes for which they were producing their art!

The Flail and Crook.

In ancient Egypt the flail, or flabellum, was known as Nekhakha, and associated primarily with the Gods Osiris and Min, while the crook was termed Heka.

As we discussed above, these royal regalia symbols are often seen being held in the hands of figurines of Osiris and on the outer coffers of the representation of the Pharaoh.

In the case of the Osiris figurines, all are usually observed in a standing position with the flail in the right hand and the crook in the left hand. In the prone Pharaoh, the majority of cases one can observe the hands crossed over the chest area, where the flail is held in the right hand and the crook in the left hand. In this case the right hand is held to the left shoulder, while the left hand is held to the right shoulder. Having said that, there are cases of the Pharaoh merely clenching each fist while bringing up the arms to rest on the chest, without actually crossing the arms. So, to all intent and purpose in this case the right hand is held to the right shoulder and the left hand held to the left shoulder.

Bearing in mind our earlier artistic explanation, it is this representation that can offer some confusion in a few cases. However, it is our finding that the flail, representing slavery, is always held in the right hand, and the crook, representing the shepherd staff leading to freedom, held in the left hand. This is meant to depict that HE is leading HIS people.

It is interesting to note that upon a Pharaoh being entombed, small wooden figurines (known as shabtis or ushabti's) are placed in the tomb with the Pharaoh. These shabti figures are also seen holding the flail in the right hand and crook in the left hand. Ushabti means "answerer", and they were placed in the tomb to spare the owner from menial corvee' labour in the after-life. Some are decorated with the text of chapter 6 of the "Book of The Dead", and thus known as the "Shabti chapter".

We now go one important step further and suggest that the flail, or symbol of slavery, held in the right hand is in fact the depiction of the mark of the beast as mentioned in Revelation. It is this "sign of man" who would endorse slavery, thus those who embraced the flail are of the "right hand" and so marking them on the right arm.

Further, and equally, the mark of the beast on the forehead relates to those whose mindset is one belief of the slavery concept.

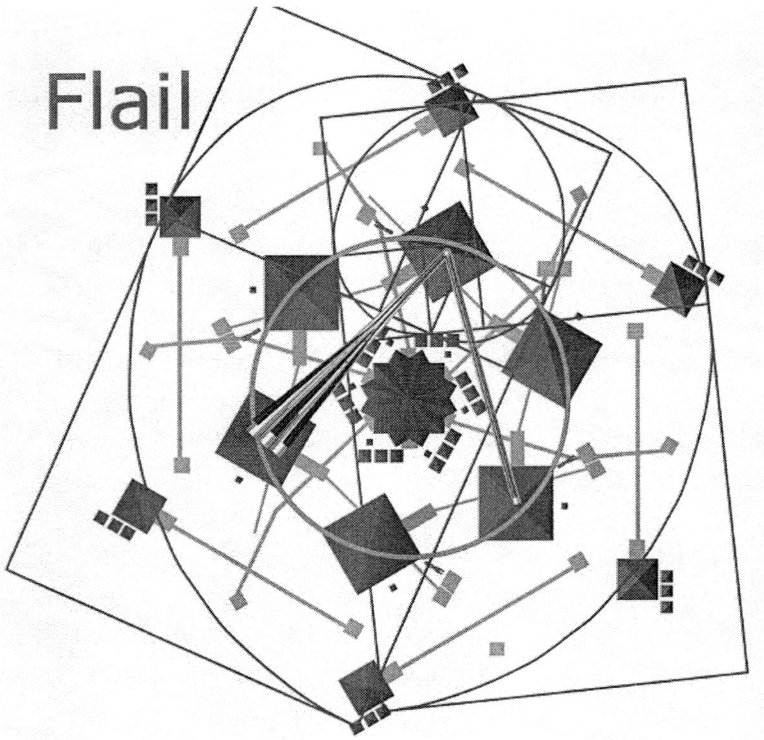

llustration 17-10 Plateau model Flail

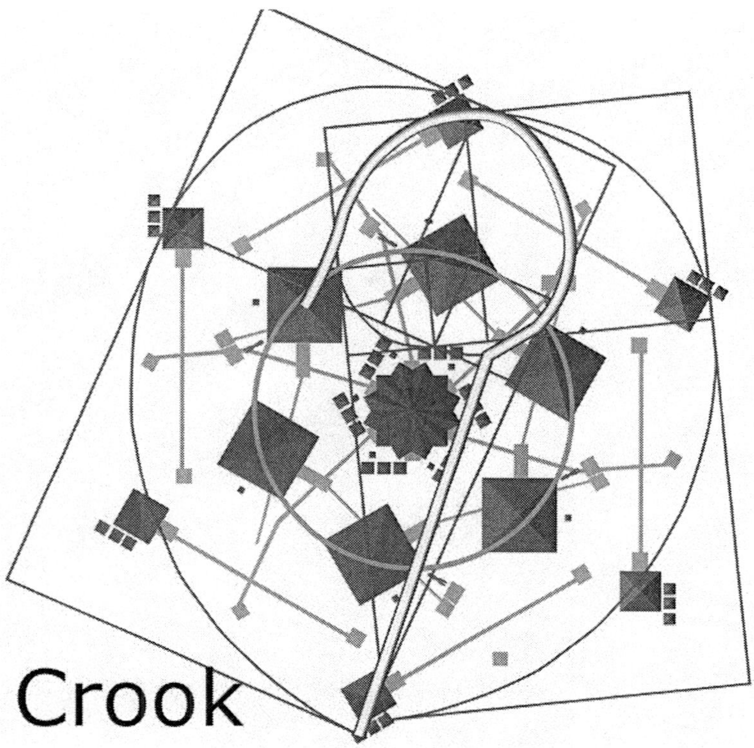

llustration 17-11 Plateau model Crook

Revelation 13:16

And he causes all, the small and the great, and the rich and the poor, and the free men and the slaves, to be given a mark on their right hand or on their forehead.

13:17

and he provides that no one will be able to buy or to sell, except the one whom has the mark, either the name of the beast or the number of his name

13:18

Here is wisdom. Let him who has understanding calculate the number of the beast, for the number is that of a man; and his number is six hundred and sixty six.

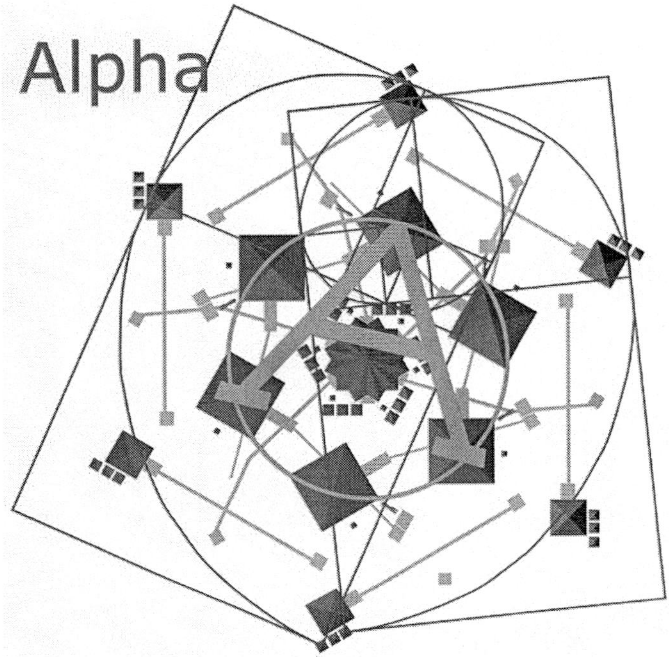

llustration 17-12 Plateau model Alpha

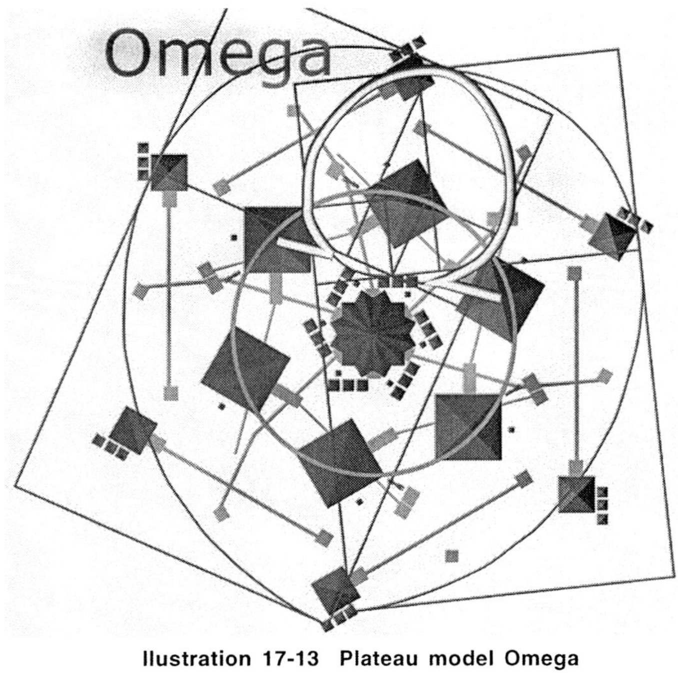

llustration 17-13 Plateau model Omega

Paternoster

In 1868 it was reported by Berem Saklatvala, in *The Christian Island*, there was found in Cirencester (England) a small piece of wall plaster from a Roman building. On this plaster were the following words making a word square.

ROTAS
OPERA
TENET
AREPO
SATOR

In 1926, Felix Grosser suggested that the above could be used in the form of a cross to write – 'Pater Noster' twice – thus

```
            A

            P
            A
            T
            E
 A PATERNOSTER O
            N
            O
            S
            T
            E
            R

            O
```

The A's and the O's are the Latin equivalent of the Greek letters ALPHA and OMEGA that were used as a symbol of Christ.

Interestingly, a similar square was discovered at Pompei and four versions at Duro – Euopos on the Euphrates. Pompeii of course was destroyed in AD 79, thus the square must be related to that year or earlier.

Notes

Latin	English
Pater	Father
Noster	Our or our /for us. So Pater noster = our father
Opera	Service or work
Sator	Sower/planter or father

In the following diagrams, the Red circle represents the closed cycle and consequently referred to as Pi. Overlaid on the Pi are the Phi sets (of which there are three). The reason for these three sets is that there are three lineages, which we have explained in detail in earlier articles.

The Queen's Pyramids located alongside (the outside edge of the model) the pyramid Menkaure give us our three points of measurement, that being the centre pyramid is our direct lineage, or our "0" line. The pyramids on either side of the centre one give us our point of measurement to which the other two lineages align. To make it two-dimensional we have drawn in the two lineage sets on the flat plane rather than at their proper angles, for the time being this is sufficient in order to visualise the information. Also, take note of the location (of each corresponding endpoint) at which each finish. For it becomes obvious that all three outer Queen's Pyramids end up aligning to the Inner Queen's Pyramids, which are found located on the plateau alongside the primary pyramid on the plateau, the Pyramid of Khufu.

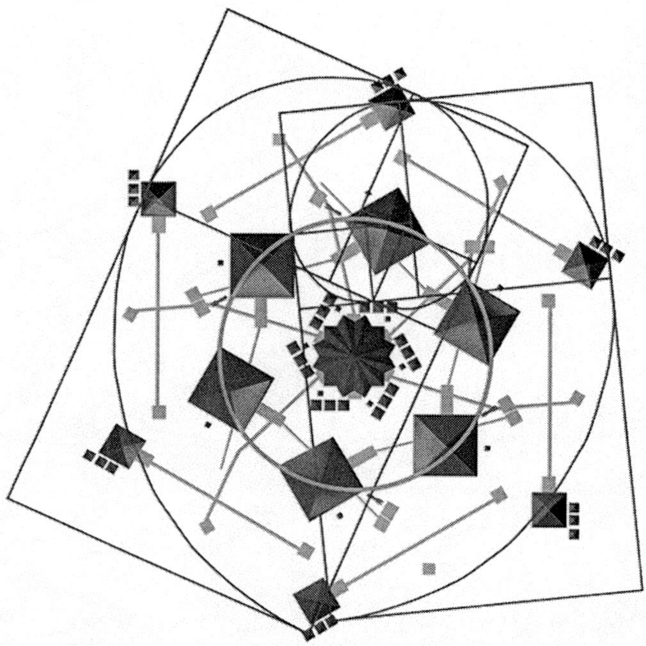

llustration 17-14 Plateau model 1

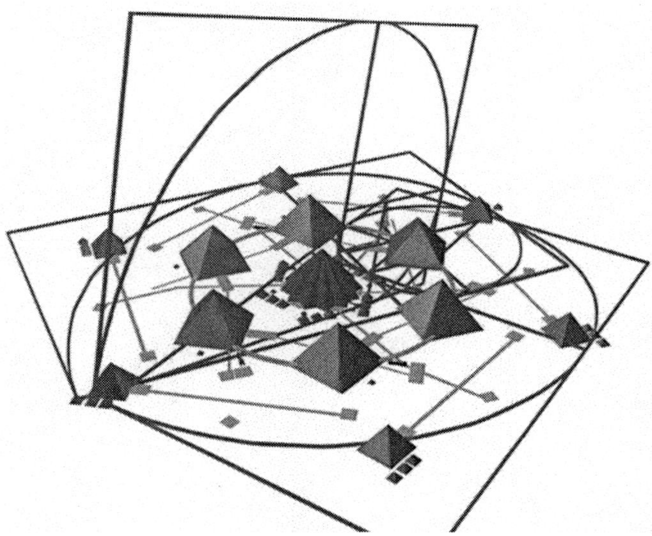

llustration 17-15 Plateau model 2

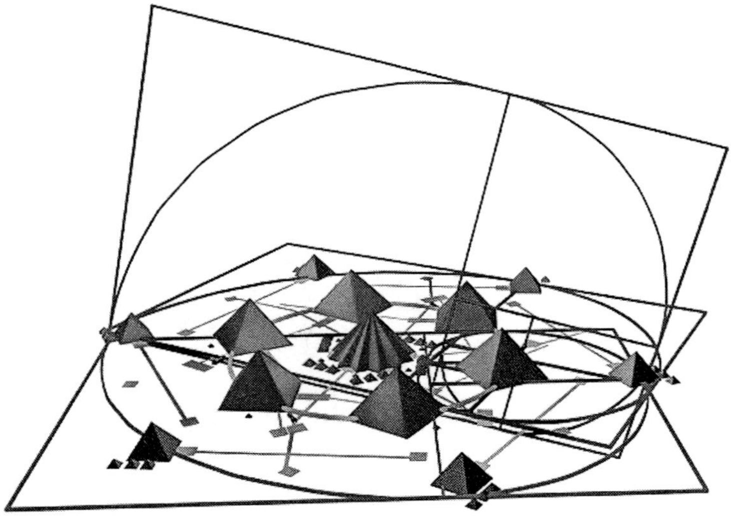

llustration 17-16 Plateau model 3

Remember that Christ was physically staked on the cross. This is to say that his physical body was put to death within the world of the lesser **Pi** or often called the **ALPHA.** But Christ is the Alpha and the Omega.

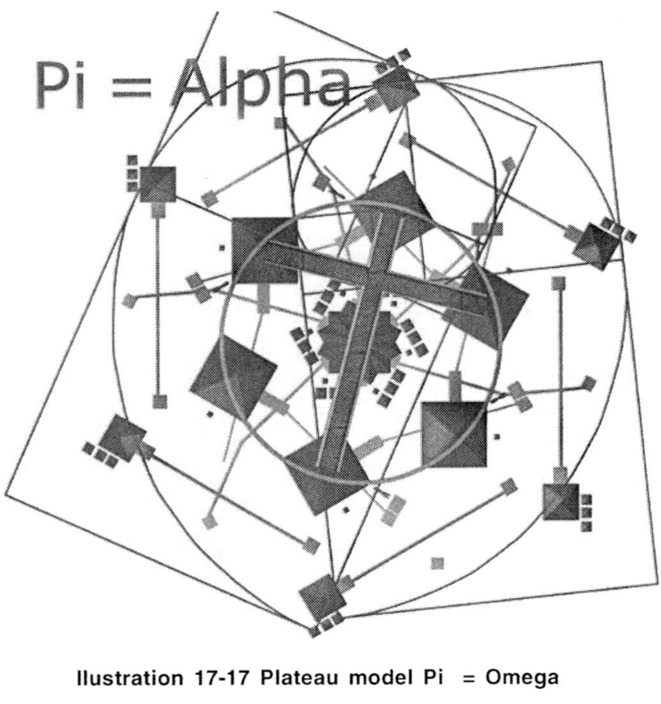

llustration 17-17 Plateau model Pi = Omega

For far greater is the life which **Phi** breaths into us, and this we call **OMEGA**

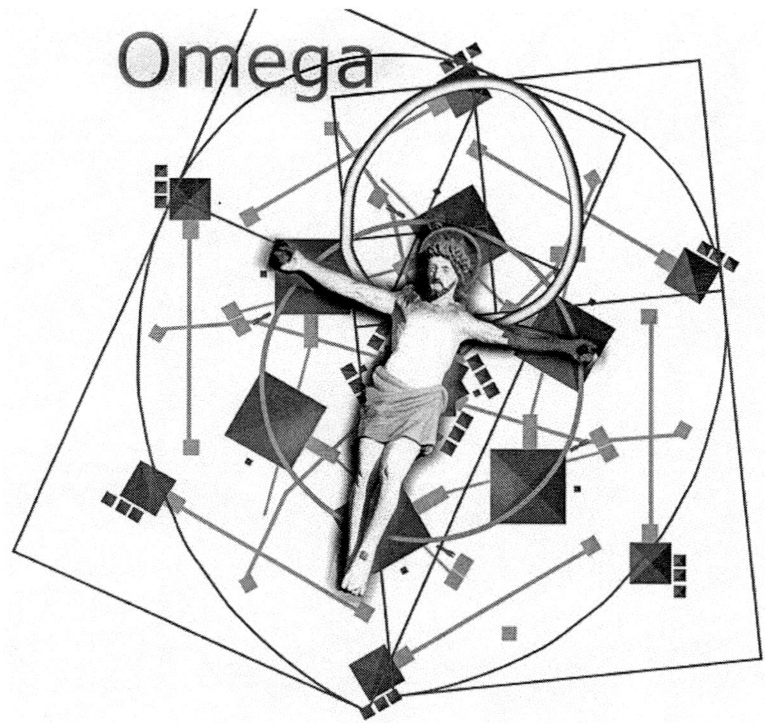

llustration 17-18 Plateau model Omega

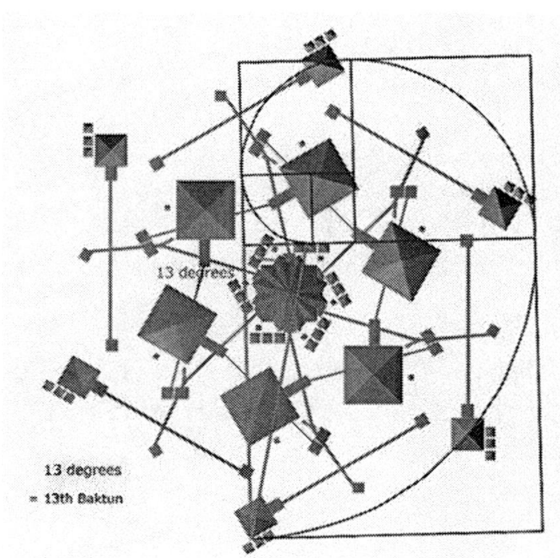

As an aide, we will briefly show how the ancient **Mayan**'s derived their **Long Count Galactic Calendar**, and how they arrived at the 13th Baktun (or the square of one).

Illustration 17-19 Plateau model and Mayan 13th Baktun

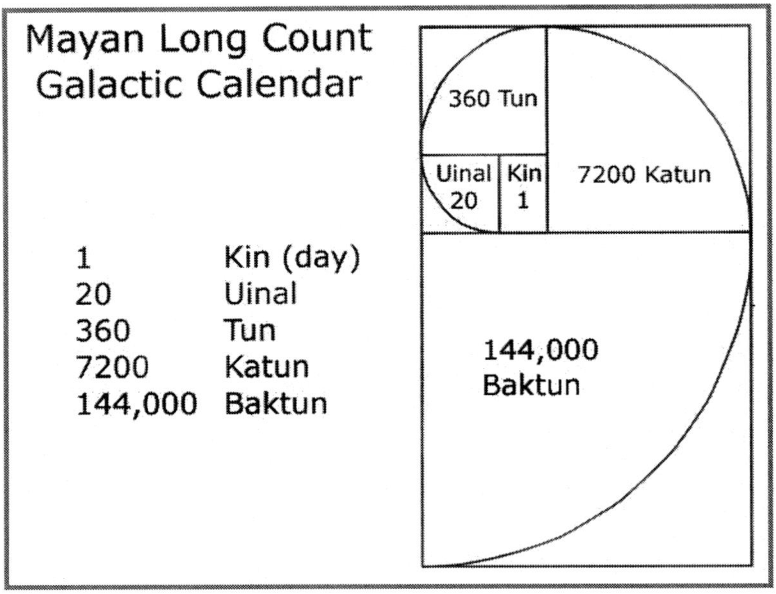

Ilustration 17-20 Mayan end count calendar diagram

BEAUTY AND THE BEAST – THE CHOICE OF LOVE

Further to our discussion of the flail and crook and weighing of the heart, let us now consider the human body. To the left of the vertical mid-point of the body is our heart - to the right is our liver. Now consider what is the purpose of each of these organs. The heart we have already explained, being the repository of our thoughts and actions, while the liver is the organ for cleaning toxins from our body. Toxins, or poisons to the human body, taken in by eating foods that are not good for our body, we can relate this also to any toxic thoughts, which in turn affects our choice of food.

So what then is the purpose of the crook? And what is the purpose of the flail? The crook guides and protects, while the flail beats and punishes. Which produces the better effect? Certainly the shepherd through his crook protects his lambs, guides them to food and safety, and the lambs grow strong through the peace provided by the shepherd and his crook. The guard however, with the flail holds his charges with fear, causing pain and loss of appetite, even loss of life, certainly loss of free will, providing no peace for his charges.

On a higher level, the left side of the human body represents the higher GOD and the right side represents the beast. We have both in each of us. By strengthening our heart, through loving thoughts and actions, we can overcome the beast in ourselves. On a mass scale, by more people choosing GOD through love, the strength multiplies geometrically, causing enough love to overcome the beast in others, who have not yet made a choice.

Many will feel fear at the events about to occur. This of course serves its purpose, in that the fear rises up, allowing a person to know there is a choice between fear or love, beast or GOD. It is at this point that many will choose. Some will choose love and GOD at this point, but the strength of the beast in many will cause the choice to be made for fear and the beast.

By choosing love now, GOD now, and overcoming the beast in ourselves before major events occur, we negate any fear that may try to rise in us. The geometrical progression of love, caused by the many making a choice prior to events occurring, overflows. This overflow of love and GOD will assist others, who have not made a choice and are fearful, to also choose love and GOD. In the case of the shepherd, the lost lamb will follow the shepherd of its own free will. However, if the lost lamb is found by the guard with the flail, it has no free will, but will try to flee the flail, and not succeeding, will be punished with pain.

The left side of the human body has often been called the "feminine side". This does not relate to gender, but to nurturing. Consider a mother and wife, ever the nurturer, just as the shepherd is to his lambs. By choosing love and GOD, we will nurture the fearful ones.

We will overcome the beast in each of us, which will disallow our beast-fear energy to be given to Esau/Lucifer. Therefore, he will no longer be able to hold us as slaves to his flail.

If someone requires sustenance, give it. If someone needs sleep, provide the means. If someone needs a ear or a shoulder, give them gladly. Through our own state of grace by choosing love and God, we will gladly help, protect and nurture those who come, which in turn will ease their fear and strengthen the energy of love, the energy which GOD requires to Free His People.

On a further scale, energy of the love we rightfully choose, the Covenant we rightfully uphold travels through the blue grid back to Giza. There it combines with everyone's love energy, fed back through the blue grid to all and everything on Earth, even Earth herself. By choosing love, choosing God, we nurture all here and we protect all here by strengthening the blue grid. Can we strengthen the grid enough through our own personal choice, which geometrically magnifies, to prevent the entry of Esau altogether? It is all up to us. What will you choose?

(Endnotes)

[1] http://www.space.com/sciencefiction/phenomena/fobos_mystery_000630.html

[2] www.danwinter.com/introduction

[3] little Pebble

Question 15: People often ask at what stage are we, and what can we expect. Set below are possible events which may take place in the next 40 years. As for the exactness of occurrence, this I cannot say other than explain what has been shown to me in Vision form, and confirmed by countless other Seers. Each episode can commingle with other events as so many happenings are to take place?

Answer 15: Let's start with the year 2000! June: a comet (Kohoutek) does a 'fly past' near Mars, causing either a collision with Mars, or creating an effect on the moon around Mars, Phobos, causing it to fly towards Earth. The moon can be shot to pieces by a rocket; thus three pieces will come to Earth, hitting France, U.S.A. and the Pacific Ocean. Apart from the main pieces there will be millions of small pieces - like a meteor shower - causing widespread fire damage and injuries. It is not clear yet if Mars itself has been hit, or its moons. Either way, the comet will cause many problems for the Earth.

Once the knowledge of this event comes to the world population, a great fear will take hold of the people, and massive anarchy and revolution will take place in cities all over the world, as people will scavenge what they can, and lawlessness will take its toll. Martial Law will be put into place. The debris of the comet, or moon, hitting the Pacific Ocean will cause tidal waves, earthquakes and volcanic eruptions, as it will cause dramatic changes to the Earth. It is to be noted that our communications will be knocked out, and a possibility is that the Olympics in Sydney may be stopped, or disrupted, due to the world events.

It is an important factor that the Seers see the dates as they are given. However, it is to be noted that historically, the Julian Calendar and all previous Christian Calendars are actually out by three to four years - e.g.: most Seers have seen the events completed by the year ending of 2000, but that date equates to 1996-97. Therefore, the 'real 2000' will be 2004; thus, all prophecies are fulfilled according to real dates.

In 1997, Hale-Bopp was thought to be the comet related to major events, but that, of course, was not the true year! - and the comet was not Hale-Bopp, but Kohoutek! Thus all events will be completed by 2004 in Julian time. Once the events concerning the Mars situation unfold - which should be around August-September, giving it approximately three months for debris to travel from Mars to Earth - the world will be in shock.

In mid-September, Pope John Paul II will make an announcement that will trigger off a major event in the Christian world, namely a new reformation - which will precipitate a division - in the Catholic Church. From this point many things will happen: civil disorder, revolution in Italy, France, England - possibly civil war in some countries.

.......A new land will rise, while the old land mass will fall.

During this whole four-year period of unrest and chastisements, some two-thirds of humanity will have died. But peace will be restored and man will rebuild the world and the Church. The old Monarchy will be restored, as democracy will have disappeared. For a while this peace will continue until the Antichrist appears, as will Enoch and Elijah. They will continue until the comet Wormwood appears and hits the Earth. The Second Coming of Jesus Christ, the Messiah, will then be apparent.

BIBLIOGRAPHY

Ambilac *Probing The Pyramids* CD-ROM series Ambilac Corporation Winnipeg 1998

Cathie, B *The Bridge to Affinity Harmonic 371244* New Zealand Quark Enterprises 1893

Edgar, J and M *The Great Pyramid Passages Vol* 1 Glasgow Bone & Hulley 1910

Edwards, IES *The Pyramids of Egypt.* New York. Viking Press 1972

Fix, Wm R *Pyramid Odyssey.* Toronto John Wiley and Sons 1978

Holy Bible *The New American Bible* Chicago The Catholic Press 1970

Lehner, M *The Complete Pyramids*, London, Thames and Hudson 1997

Lemusurier, P *Great Pyramid decoded.* Dorset, Element 1996

Middleton-Jones, H *Giza-Genesis, The Best Kept Secrets. Vol 1* Arizona. Galileo Publishing

Wilkie, M J Books 2001

Seton-Williams *Blue Guide to Egypt*, 1st edition. London. Ernest Benn Ltd 1983

Smith, P *The Great Pyramid.* New York. Random House 1978

Tompkins , P *Secrets of the Great Pyramid*, New York, Harper and Row 1978

The Authors

Howard Middleton-Jones

An author,investigative archaeologist, Egyptologist and frequent visitor to Egypt, Howard is well-known on the conference circuit, presenting talks on the mysteries of our ancient origins. He was born in South Wales, UK, and received his education at the University of Wales, Swansea, and Oxford University. Howard has two grown up daughters, and now shares his life with his philosopher partner, Jenni. They live in the old sea fishing village of Mumbles near Swansea, South Wales.

James M. Wilkie

For over 20 years, James has sought information concerning the secrets contained in the pyramids at Giza. Through meticulous research and global travel, he continued to discover numerous sources that appeared to be pointing him in a specific direction. At length, he was able to correlate and analyze this information, leading him to the conclusions that are presented in this book. James lives in Winnipeg, Manitoba Canada, and maintains the following website: www.ambilac.com

Index

Symbols

A

B

C

D

E

F

G

H

I

J

K

L

M

N

O

P

Q

R

S

T

U

V

W

X

Y

Z